SpringerBriefs in Computer Science

Series Editors
Stan Zdonik
Peng Ning
Shashi Shekhar
Jonathan Katz
Xindong Wu
Lakhmi C. Jain
David Padua
Xuemin (Sherman) Shen
Borko Furht
V.S. Subrahmanian
Martial Hebert
Katsushi Ikeuchi
Bruno Siciliano

For further volumes:
http://www.springer.com/series/10028

Hongwei Li

Enabling Secure and Privacy Preserving Communications in Smart Grids

 Springer

Hongwei Li
University of Electronic Science
   and Technology of China
Chengdu, Sichuan
People's Republic of China

ISSN 2191-5768   ISSN 2191-5776 (electronic)
ISBN 978-3-319-04944-1   ISBN 978-3-319-04945-8 (eBook)
DOI 10.1007/978-3-319-04945-8
Springer Cham Heidelberg New York Dordrecht London

Library of Congress Control Number: 2014933524

Printed on acid-free paper

Springer is part of Springer Science+Business Media (www.springer.com)

Preface

A smart grid has emerged as a promising solution to the next generation power grid system. It utilizes information and communications technology to gather and act on information, such as the behavior of suppliers and consumers, in an automated fashion to improve the reliability, efficiency, economics, and sustainability of the generation and distribution of electricity. However, security and privacy issues still present practical concerns to the deployment of smart grids. In this book, we investigate three schemes for secure and privacy-preserving smart grid communications.

In Chap. 2, we present an efficient privacy-preserving demand response scheme which employs a homomorphic encryption to achieve privacy-preserving demand aggregation and efficient response. In addition, an adaptive key evolution technique is further investigated to ensure the users' session keys to be forward secure. In Chap. 3, we introduce an efficient authentication scheme which utilizes the Merkle hash tree technique to secure smart grid communication. Specifically, the proposed authentication scheme considers the smart meters with computation-constrained resources and puts the minimum computation overhead on them. In Chap. 4, an efficient fine-grained keywords comparison scheme is proposed. Based on the homomorphic Pailier cryptosystem, we use two super-increasing sequences to aggregate multidimensional keywords. As a result, the comparison between the keywords of all sellers and those of one buyer can be achieved with only one calculation.

This book presents an overview of the state-of-the-art solutions to secure and privacy-preserving communications in smart grids. It not only reveals unique security and privacy characteristics but also offers effective solutions. Security analysis and performance evaluation demonstrate effectiveness and efficiency of three schemes. Last but not least, this book highlights promising future research directions to guide interested readers.

Sichuan, China   Hongwei Li

Acknowledgments

This book is supported by the National Natural Science Foundation of China under Grants 61350110238, 61103207, U1233108, 61272525, 61073106, and 61003232; the Fundamental Research Funds for Chinese Central Universities under Grant ZYGX2011J059; and the 2011 Korea-China Young Scientist Exchange Program.

Acknowledgements

Contents

Chapter 1
Introduction to Smart Grids

1.1 Smart Grids

Lack of effective real-time diagnosis and healing, the traditional power grid is sporadically suffering from failures and blackouts. For example, on August 14, 2003, power system outage affected large portions of the north eastern U.S. and Canada, which ultimately caused a $6 billion loss in economic revenue [1]. Recently, smart grids have emerged as a promising solution to the next generation power grid system [2]. It utilizes information and communications technology to gather and act on information, such as the behavior of suppliers and consumers in an automated fashion to improve the reliability, efficiency, economics, and sustainability of the generation and distribution of electricity [3].

1.1.1 Communication Network Architecture

As shown in Fig. 1.1 [4, 5], a smart grid consists of seven logical domains:

- Bulk Generation: Bulk generation produces electric power by different means such as hydropower, solar, wind, tidal forces, and other generation sources.
- Transmission: A very high voltage infrastructure transfers electrical energy from power plants to electrical substations.
- Distribution: Distribution networks step down voltage and deliver electricity from substations to consumers.
- Customer: A consumer uses the electric energy in a multitude of ways and pays for electrical goods or services.
- Markets: The balance between the supply and the demand of electricity is maintained by the markets, which consist of retailers who supply electricity to end users, suppliers of bulk electricity, traders who buy electricity from suppliers and sell it to retailers.

H. Li, *Enabling Secure and Privacy Preserving Communications in Smart Grids*,
SpringerBriefs in Computer Science, DOI 10.1007/978-3-319-04945-8_1,
© The Author(s) 2014

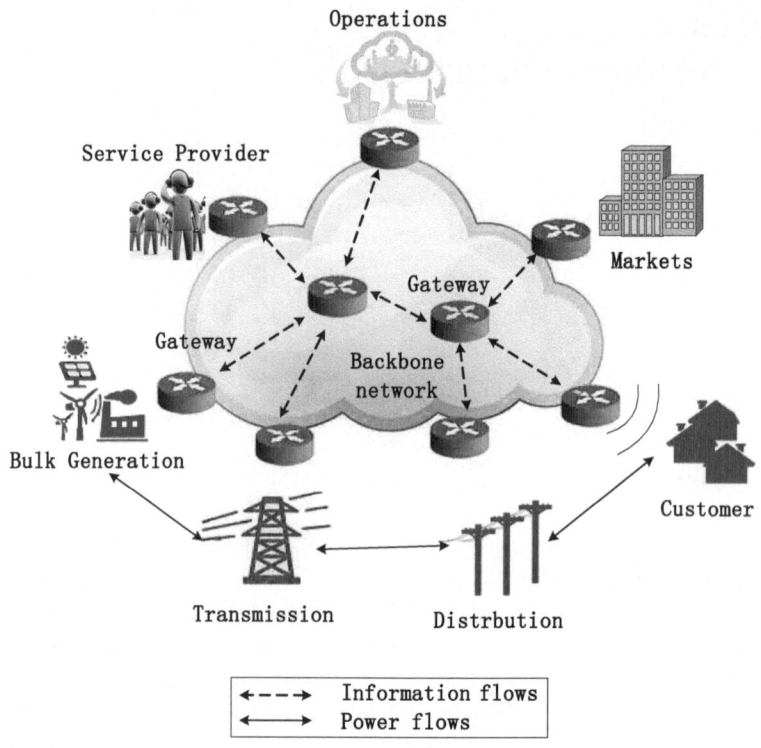

Fig. 1.1 Communication network architecture in a smart grid

- Service Provider: The service provider performs services to support the business processes of power system generators and consumers.
- Operations: Operations use field area and wide area networks in the transmission and distribution domains to obtain information of power system activities like monitoring, control, fault management, maintenance, analysis and metering.

The first four feature the two-way power and information flows. The last three feature information collection and power management in a smart grid. In order to interconnect all these domains, the communication network must be highly distributed and hierarchical.

1.1.2   Characteristics of Smart Grids

1.1.2.1   Reliability

Smart grids make use of technologies that improve fault detection and allow self-healing of the network without the intervention of technicians. This ensures a more reliable supply of electricity, and reduces vulnerability to natural disasters or attack. Although multiple routes are touted as a feature of smart grids, the traditional electricity grid also featured multiple routes. Initial power lines in the grid were built using a radial model; later connectivity is guaranteed via multiple routes, referred to as a network structure. However, that created a new problem: if the current flow or related effects across the network exceed the limits of any particular network element, it could fail, and the current would be shunted to other network elements, which eventually may fail also, causing a domino effect [6].

1.1.2.2 Market-Enabling

Smart grids allow for systematic communication between suppliers (their energy price) and consumers (their willingness-to-pay), and permits both the suppliers and the consumers to be more flexible and sophisticated in their operational strategies. Only the critical loads need to pay the peak energy prices, and consumers are able to be more strategic when they use energy. Generators with greater flexibility are able to sell energy strategically for maximum profit, whereas inflexible generators such as base-load steam turbines and wind turbines receive a varying tariff based on the level of demand and the status of the other generators currently operating. The overall effect is a signal that awards energy efficiency, and energy consumption that is sensitive to the time-varying limitations of the supply. At the domestic level, appliances with a degree of energy storage or thermal mass (such as refrigerators, heat banks, and heat pumps) are well-placed to play the market and seek to minimize energy cost by adapting demand to the lower-cost energy support periods. This is an extension of the dual-tariff energy pricing mentioned above [6].

1.1.2.3 Demand Response

Demand response (DR) could be an effective alternative to deal with the increasing demand for electricity. Because DR works directly on the customer side of the meter, it displaces both generation and transmission. A well-designed DR scheme can effectively and efficiently yield peak shaving capability that could alleviate the shortage occurring under the emergent circumstances [7]. DR is generally used to refer to mechanisms used to encourage consumers to reduce demand, thereby reducing the peak demand for electricity. Since electrical generation and transmission systems are generally sized to correspond to peak demand, lowering peak demand reduces overall plant and capital cost requirements. Depending on the configuration of generation capacity, however, DR may also be used to increase demand load at times of high production and low demand. Some systems may thereby encourage energy storage to arbitrage between periods of low and high demand (or low and high prices) [8].

1.1.2.4 Security and Privacy

Security-related research is very critical to the success of smart grids. A smart grid is a promising power delivery infrastructure integrated with communication and information technologies. The increased interconnection and integration also introduce cyber vulnerabilities into the grid [9]. Security attackers can easily approach the cyber-physical space of smart grids and launch various attacks, such as data modification attacks and data injection attacks.

Data modification attacks are typical attacks in smart grids. Such attacks attempt to stealthily modify data in order to corrupt critical information exchange in smart grids. The target can be either customers' information (e.g., pricing information and account balance) or status values of power systems (e.g., voltage readings and device running status). Because such information in power systems is valuable to both end users and utility companies, fault-tolerant and integrity-check methods are deployed in power systems to protect data integrity. However, the risk of integrity attacks is indeed real [10]. Data injection attacks are notorious attacks in smart grids. An adversary can send packets to inject false information on current or future prices, or send wrong meter data to a utility company. The result of injecting false prices, such as negative pricing, will be power shortage or other significant damages on the target region. The result of sending wrong data include reduced electric bills for economic damages due to the loss of revenue of a utility company [11]. Privacy preservation is another significant research issue in smart grids [9]. Intelligent control and economic management of energy consumption require more interoperability between consumers and service providers. Unprotected energy-related data will cause invasions of privacy in smart grids. In particular, radio waves in AMI may disclose information about where people were and when and what they were doing. Failure to address privacy issues in smart grids will not be accepted by regulators and customers.

1.2 Research Topics in Smart Grids

In this book, we will study three research topics that are related to the security and privacy in smart grids. Specifically, we will design novel communication protocols and show that smart grid characteristics can be applied to improve the security and efficiency of communication protocols.

Privacy-Preserving Demand Response: Firstly, we study the demand response problem. Demand response (DR), which can assist users to use energy efficiently and transfer non-emergent power demand from on-peak time to off-peak time [12]. DR can also bring various benefits to users. For example, users can reduce their electricity expenditure by matching the operation time of different electric appliances in their places to the period with the cheapest price. As smart grids are closely related to people's daily lives, to resolve the security and privacy concerns in smart grids is crucial [13, 14]. Note that in smart grid communications, adversaries might eavesdrop on the communication between the users and the control center as well as identify the users' electricity demand. With this information, they are able to track and learn about the users' habits or lifestyles [15]. Moreover, adversaries might compromise the smart meters and further obtain stored secret information such as their session keys and private keys [16]. To preserve users' privacy and cyber security, DR should not only provide privacy preservation of electricity demand, but also mitigate the damage caused by the exposure of secret keys stored on the smart meters. In this book, we review some related schemes and study the security goals. We find out that existing schemes mainly focus on achieving confidentiality and integrity of communication, and mutual authentication among different entities [17, 18]. The first attempt to achieve forward secrecy of users' session keys in smart grids is studied in [19]. However, since it adopts RSA public key algorithm and Diffie-Hellman exchange protocol to evolve the session key as an effort to ensure forward secrecy, the computation and communication overhead are heavy, thereby making it impractical. Thus, we propose an efficient privacy-preserving demand response (EPPDR) scheme with adaptive key evolution. It realizes secure and efficient electricity demand aggregation and response based on the homomorphic encryption and the key evolution techniques. In addition, EPPDR can adaptively control the key evolution to balance the trade-off between the communication efficiency and security level [20].

Lightweight Authentication: Secondly, we study the lightweight authentication in smart grids. Smart grids provide an attractive feature, i.e., two-way information flow communication, in which the neighborhood gateway can collect electricity consumption reports from the customers via a wireless connection. Then, the neighborhood gateway sends the electricity reports to the control center via a wired link with high bandwidth and low delay. Based on the statistics and analysis of the above electricity reports, the control center can further convey the real-time pricing information to customers for lower electricity bills, or send the control information to flatten demand peak [15, 21]. The consumption-reporting device at the customer side is called the smart meter, which is vulnerable to malicious operations, e.g., the meter's reading modification [22]. It is indispensable for electricity utility to develop an authentication scheme to prevent malicious operations. In addition, a smart meter is only equipped with limited resources, i.e., a computation-constrained microprocessor, a small memory and a low computational capacity, etc. However, the computation overhead is heavy for the smart meter. Thus, the developed authentication scheme should minimize computation overhead on the smart meters. In this book, we will propose a novel authentication scheme, where the Merkle hash tree technique is leveraged to facilitate the authentication implementation. The security analysis indicates that the proposed scheme can resist the replay attack, the message injection attack, the message analysis attack and the message modification attack. In addition, performance evaluation demonstrates that the proposed authentication scheme can achieve less communication overhead and dramatically reduce computation cost compared with the traditional authentication scheme, e.g., RSA-based authentication [23].

Fine-Grained Keywords Comparison: Thirdly, we study the range query in the smart grid auction market. The energy auction market introduces commercial auctions to smart grids, where energy sellers publish their auction information, and then energy buyers bid for appropriate energy supplies. Thus, the energy auction market can adjust energy prices and provide strong support for the practical application of smart grids [24]. Unfortunately, security and privacy are seriously challenged in the energy auction market. Firstly, due to the particularity of the energy auction market, privacy preservation is extremely important because auction information is closely related to trade secrets. In addition, many traditional security requirements are still needed: sensitive information should be encrypted, legitimate users should be authenticated, and illegal users cannot modify communication messages. In this book, we will propose an efficient fine-grained keywords comparison (EFKC) scheme in the smart grid auction market. This scheme focuses on providing secure and efficient transactions between generators and retailers. It considers auction message encryption, keyword search, fine-grained comparison and auction message pre-filtering. Security analysis demonstrates that EFKC can achieve privacy preservation, authentication, data integrity and confidentiality. Performance evaluation shows that EFKC significantly improves computation and communication efficiency compared with the existing scheme in [25].

1.3 Security Primitives

In this section, we will review some basic techniques.

1.3.1 Homomorphic Encryption

Homomorphic Encryption (HE) allows certain algebraic operations on the plaintext to be performed directly on the ciphertext. HE is usually used for privacy-preserving applications (e.g., data aggregation, e-voting). In this chapter, we adopt the Paillier cryptosystem [26]. In the Paillier cryptosystem, the public key is $pk(N, g)$, and the corresponding private key is $sk(\lambda, \mu)$. Let $E(\cdot)$, m, and r be the encryption function, a message and a random number, respectively. The ciphertext is

$$c = E(m) = g^m \cdot r^N \bmod N^2 \tag{1.1}$$

The plaintext is

$$m = D(c) = L(c^{\lambda \bmod N^2}) \cdot \mu \bmod N \tag{1.2}$$

where the function $L(x) = (x - 1)/N$. Then, the additive homomorphic property is as follows:

$$E(m_1) \cdot E(m_2) = (g^{m_1} \cdot r_1^N)(g^{m_2} \cdot r_2^N) \bmod N^2$$
$$= g^{m_1+m_2} \cdot (r_1 r_2)^N \bmod N^2 \qquad (1.3)$$
$$= E(m_1 + m_2)$$

1.3.2 Bilinear Pairing

Let $\mathbb{G}$ and $\mathbb{G}_T$ be two multiplicative cyclic groups of the same prime order q, and P be a generator of group $\mathbb{G}$. Suppose $\mathbb{G}$ and $\mathbb{G}_T$ are equipped with a pairing, i.e., a non-degenerated and efficiently computable bilinear map $e : \mathbb{G} \times \mathbb{G} \to \mathbb{G}_T$ such that $e(aP_1, bQ_1) = e(P_1, Q_1)^{ab} \in \mathbb{G}_T$ for all $a, b \in \mathbb{Z}_q^*$ and any $P_1, Q_1 \in \mathbb{G}$. We refer to [27] for a more comprehensive description of the pairing technique and complexity assumptions.

Definition 1.1. A bilinear parameter generator $\mathscr{G}en$ is a probabilistic algorithm that takes a security parameter κ as input, and outputs a 5-tuple $(q, P, \mathbb{G}, \mathbb{G}_T, e)$.

1.3.3 Identity-Based Signature

Identity-based signature is made of four algorithms that are depicted as follows [28]:

- Setup: Private key generator (PKG) firstly generates $(q, P, \mathbb{G}, \mathbb{G}_T, e)$ by running $\mathscr{G}en(\kappa)$. Then PKG chooses a random $s \in \mathbb{Z}_q^*$ as the master key and computes the associated public key $P_{pub} = sP$. It also picks two cryptographic hash functions of the same domain and range $H_1, H_2 : \{0, 1\}^* \to \mathbb{G}$. The system's public parameters are $(q, P, \mathbb{G}, \mathbb{G}_T, e, H_1, H_2)$.
- Keygen: Given a user's identity ID, PKG computes $Q_{ID} = H_1(ID)$ and the associated private key $d_{ID} = sQ_{ID}$ that is transmitted to the user.
- Sign: In order to sign message M, the user picks a random number $r \in \mathbb{Z}_q^*$, and computes $U = rP$, $V = d_{ID} + rH_2(ID, M, U)$. The signature on M is the pair $\sigma = <U, V>$.
- Verify: To verify a signature $\sigma = <U, V>$ on a message M for an identity ID, the verifier accepts the signature if $e(P, V) = e(P_{pub}, H_1(ID))e(U, H_2(ID, M, U))$ and rejects it otherwise.

1.3.4 Merkle Hash Tree

The main idea of the Merkle hash tree is to construct a tree based on a one-way cryptographic hash function $h(.)$ [29], then each leaf node can be verified through its Authentication Path Information (API). Since only the hash functions are computed, the computation cost of verification is very low. We illustrate the construction and application of the Merkle hash tree through an example. As shown in Fig. 1.2, the values of the eight leaf nodes are the message hashes, $h_i = h(D_i)(i = 1, \cdots, 8)$, respectively. The values of internal nodes are derived from their child nodes. For instance, the value of the node $N_{3,4}$ is $h_{3,4} = h(h_3|h_4)$ and the value of the root node $N_{1,8}$ is $h_{1,8} = h(h_{1,4}|h_{5,8})$. Each leaf node can be verified with $h_{1,8}$ and the corresponding API. For instance, the node N_1 can be authenticated by the server who stores $h_{1,8}$ as follows. N_1 sends D_1 and the corresponding API $= < h_2, h_{3,4}, h_{5,8} >$ to the server. Then the server can check the authenticity of node N_1 by firstly computing $h_1, h_{1,2} = h(h_1|h_2), h_{1,4} = h(h_{1,2}|h_{3,4}), h_{1,8} = h(h_{1,4}|h_{5,8})$. And then, the server checks whether the computed $h_{1,8}$ is the same as the existing $h_{1,8}$. The server accepts N_1, only if the two values are equal.

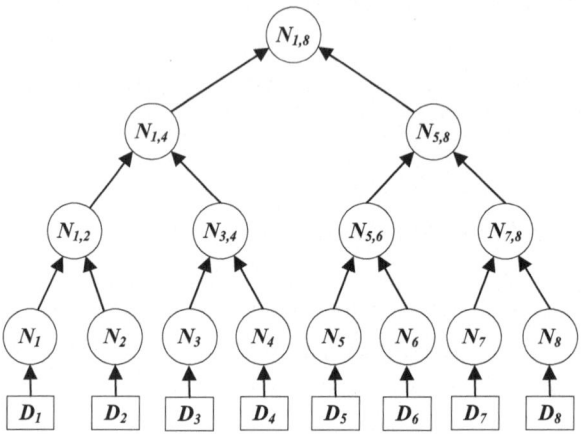

Fig. 1.2 Merkle hash tree

1.4 Summary

In this chapter, we provide an overview of smart grids such as communication network architecture, characteristics, and research topics in smart grids. In the following chapters, we present three state-of-the-art solutions to secure privacy-preserving communications in smart grids. The three representative schemes explore the fundamental issues such as privacy-preserving demand response, lightweight authentication, and fine-grained keywords comparison in a comprehensive fashion.

References

1. "blackout 2003," http://www.ieso.ca/imoweb/EmergencyPrep/blackout2003.
2. H. Liang, B. Choi, W. Zhuang, and X. Shen, "Towards optimal energy store-carry-and-deliver for phevs via v2g system," in *IEEE INFOCOM*, 2012, pp. 1674–1682.
3. H. Liang, B. Choi, A. Abdrabou, W. Zhuang, and X. Shen, "Decentralized economic dispatch in microgrids via heterogeneous wireless networks," *IEEE Journal on Selected Areas in Communications*, vol. 30, no. 6, pp. 1061–1074, 2012.
4. "Nist framework and roadmap for smart grid interoperability standards," *Release 1.0, NIST Special Publication 1108*, pp. 1–145, 2010.
5. D. He, C. Chen, J. Bu, S. Chan, Y. Zhang, and M. Guizani, "Secure service provision in smart grid communications," *IEEE Communications Magazine*, vol. 50, no. 8, pp. 53–61, 2012.
6. "Smart grid," http://en.wikipedia.org/wiki/Smart_grid.
7. J. Wang, C. N. Bloyd, Z. Hu, and Z. Tan, "Demand response in China," *Energy*, vol. 35, no. 4, pp. 1592–1597, 2010.
8. "Demand response," http://en.wikipedia.org/wiki/Demand_response.
9. J. Liu, Y. Xiao, S. Li, W. Liang, and C. Chen, "Cyber security and privacy issues in smart grids," *IEEE Communications Surveys & Tutorials*, vol. 14, no. 4, pp. 981–997, 2012.
10. W. Wang and Z. Lu, "Cyber security in the smart grid: Survey and challenges," *Computer Networks*, vol. 57, no. 5, pp. 1344–1371, 2013.
11. Y. Mo, T.-J. Kim, K. Brancik, D. Dickinson, H. Lee, A. Perrig, and B. Sinopoli, "Cyber–physical security of a smart grid infrastructure," *Proceedings of the IEEE*, vol. 100, no. 1, pp. 195–209, 2012.
12. F. Rahimi and A. Ipakchi, "Demand response as a market resource under the smart grid paradigm," *IEEE Transactions on Smart Grid*, vol. 1, no. 1, pp. 82–88, 2010.
13. X. Li, X. Liang, R. Lu, H. Zhu, X. Lin, and X. Shen, "Securing smart grid: Cyber attacks, countermeasures and challenges," *IEEE Communications Magazine*, vol. 58, no. 8, pp. 38–45, 2012.
14. X. Liang, X. Li, R. Lu, X. Lin, and X. Shen, "Udp: Usage-based dynamic pricing with privacy preservation for smart grid," *IEEE Transactions on Smart Grid*, vol. 4, no. 1, pp. 141–150, 2013.
15. H. Li, X. Liang, R. Lu, X. Lin, and X. Shen, "Edr: An efficient demand response scheme for achieving forward secrecy in smart grid," in *IEEE GLOBECOM*, 2012, pp. 929–934.
16. H. Li, R. Lu, L. Zhou, B. Yang, and X. Shen, "An efficient merkle tree based authentication scheme for smart grid," *IEEE Systems Journal*, http://ieeexplore.ieee.org/xpls/abs_all.jsp?arnumber=6563123.
17. F. Li, B. Luo, and P. Liu, "Secure information aggregation for smart grids using homomorphic encryption," in *2010 IEEE International Conference on Smart Grid Communications (Smart-GridComm)*, 2010, pp. 327–332.
18. R. Lu, X. Liang, X. Li, X. Lin, and X. Shen, "Eppa: An efficient and privacy preserving aggregation scheme for secure smart grid communications," *IEEE Transactions on Parallel and Distributed Systems*, vol. 23, no. 9, pp. 1621–1631, 2012.
19. M. Fouda, Z. Fadlullah, N. Kato, R. Lu, and X. Shen, "A lightweight message authentication scheme for smart grid communications," *IEEE Transactions on Smart Grid*, vol. 2, no. 4, pp. 675–685, 2011.
20. H. Li, X. Lin, H. Yang, X. Liang, R. Lu, and X. Shen, "Eppdr: An efficient privacy-preserving demand response scheme with adaptive key evolution in smart grid," *IEEE Transactions on Parallel and Distributed Systems,*, http://www.computer.org/csdl/trans/td/preprint/06506075-abs.html.
21. H. Liu, H. Ning, Y. Zhang, and L. Yang, "Aggregated-proofs based privacy-preserving authentication for v2g networks in smart grid," *IEEE Transactions on Smart Grid*, vol. 3, no. 4, pp. 1722–1733, 2012.

22. M. Wen, R. Lu, K. Zhang, J. Lei, X. Liang, and X. Shen, "Parq: A privacy- preserving range query scheme over encrypted metering data for smart grid," *IEEE Transactions on Emerging Topics in Computing*, vol. 1, no. 1, pp. 178–191, 2013.
23. "Introduction to nistir 7628 guidelines for smart grid cyber security," *Available:* http://www. nist.gov/smartgrid/upload//nistir-7628$_-$total.pdf, 2010.
24. D. Kang, B. Kim, and D. Hur, "Supplier bidding strategy based on non-cooperative game theory concepts in single auction power pools," *Electric power systems research*, vol. 77, no. 5, pp. 630–636, 2007.
25. M. Wen, R. Lu, J. Lei, H. Li, X. Liang, and X. Shen, "Sesa:an efficient searchable encryption scheme for auction in emerging smart grid marketing," *Security and Communication Networks,* http://onlinelibrary.wiley.com/doi/10.1002/sec.699/full, 2013.
26. P. Paillier, "Public-key cryptosystems based on composite degree residuosity classes," in *EUROCRYPT*, 1999, pp. 223–238.
27. D. Boneh and M. K. Franklin, "Identity-based encryption from the weil pairing," in *CRYPTO*, 2001, pp. 213–229.
28. B. Libert and J. Quisquater, "The exact security of an identity based signature and its applications," *Preprint available at* http://eprint.iacr.org/2004/102, 2004.
29. R. Rivest, "Rfc 1321: The md5 message-digest algorithm," *Internet activities board*, vol. 143, 1992.

Chapter 2
Privacy-Preserving Demand Response in Smart Grids

2.1 Introduction

Recently, smart grids have attracted increasing attention [1–4]. Compared with the traditional power grid, smart grids are featured with many attractive characteristics, e.g., self-monitoring, self-healing, remote check, pervasive control and more customer choices [5–8]. One appealing feature of smart grids is demand response (DR), which can assist users to use energy efficiently and transfer non-emergent power demand from on-peak time to off-peak time [9]. DR can also bring various benefits to users. For example, users can reduce their electricity expenditure by matching the operation time of different electric appliances in their places to the period with the cheapest price.

Since smart grids are closely related to people's daily lives, to resolve the security and privacy concerns in smart grids is crucial [10, 11]. Note that adversaries might compromise the smart meters and further obtain stored secret information such as their session keys and private keys [12]. Moreover, in smart grid networks, adversaries might eavesdrop on the communication between the users and the control center and identify the users' electricity demand. With this information, they are able to track and learn about the users' habits or lifestyles [13]. To preserve user privacy and cyber security, DR should not only provide privacy preservation of electricity demand, but also mitigate the damage caused by the exposure of secret keys stored on the smart meters. Among many security and privacy requirements for protection of electricity demand and response messages in smart grids, forward secrecy is extremely important since cryptographic computations, e.g., encryption, signature and authentication, are often carried out on the insecure smart meters [14]. In a scheme with forward secrecy, secret keys are evolved at regular time periods. Exposure of a secret key corresponding to a given time period does not enable an adversary to break the scheme for any prior time period [15]. To improve the security level of smart meters, forward secrecy should be considered. Despite its importance, forward secrecy has not been studied well in smart grids due to the complexity of smart grid communication. Existing schemes mainly focus on

H. Li, *Enabling Secure and Privacy Preserving Communications in Smart Grids*,
SpringerBriefs in Computer Science, DOI 10.1007/978-3-319-04945-8_2,
© The Author(s) 2014

achieving confidentiality and integrity of communication, and mutual authentication among different entities [3, 16]. The first attempt to achieve forward secrecy of users' session keys in smart grids was studied in [17]. However, since it adopts RSA public key algorithm and Diffie-Hellman exchange protocol to evolve the session key as an effort to ensure forward secrecy, the computation and communication overheads are heavy, thereby making it impractical. In addition, the private keys of users should be evolved since they could also be compromised [14]. However, frequently evolving session keys and private keys will lead to heavy communication overhead; nevertheless, sparsely evolving session keys and private keys will degrade the security level. Therefore, it is challenging to develop a key evolution algorithm that can achieve both efficiency and security levels.

In this chapter, we propose an efficient Privacy-preserving Demand Response scheme with adaptive key evolution, named PDR. Specifically, the contributions of this chapter are twofold. Firstly, we propose the novel PDR scheme that employs the homomorphic encryption [18] to achieve privacy-preserving demand aggregation and efficient response. The security analysis demonstrates that PDR can achieve privacy preservation of electricity demand, forward secrecy of users' session keys, and evolution of users' private keys. Secondly, we compare PDR with an existing scheme [17] which also achieves forward secrecy. The comparison results demonstrate that PDR is more efficient in terms of computation and communication overheads.

The remainder of the chapter is organized as follows. In Sect. 2.2, we formalize the network model, the security model and the design goal. Then, we propose the PDR scheme in Sect. 2.3, followed by the security analysis and the performance evaluation in Sect. 2.4 and Sect. 2.5, respectively. We present the related works in Sect. 2.6. Finally, we draw our summary in Sect. 2.7.

2.2 Models and Design Goal

2.2.1 Network Model

As shown in Fig. 2.1, the network model for a smart grid is divided into a number of hierarchical networks comprising of control center (CC), building area network (BAN), and home area network (HAN). The CC covers n BANs. For the sake of simplicity, we assume each BAN comprises m HANs. Each HAN is assigned a smart meter enabling an automated, two-way communication between the CC and the HAN user. Meantime, each BAN is equipped with a gateway (GW). Communication between a HAN user and the BAN GW (BG) is through relatively inexpensive WiFi technology, i.e., within the WiFi coverage of the BG, each HAN user can directly communicate with the BG. For the HAN user who is beyond the coverage of the BG, communication has to be done in multiple hops. However, since the distance between a BG and the CC is far away, the communication between a BG and the CC is through either wired links or any other links with high bandwidth and low delay.

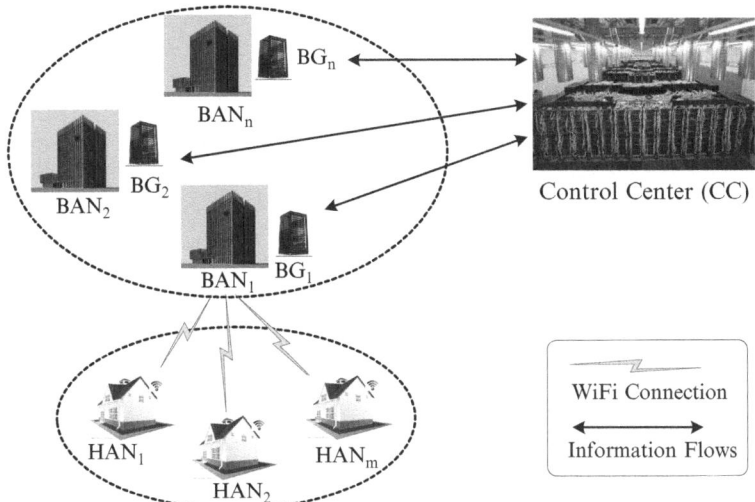

Fig. 2.1   Network model for a smart grid

2.2.2   Security Model

In the security model, CC and BGs are trusted by all parties in the scheme, and infeasible for any adversary to compromise. In specific, we consider the following security goals needed to be achieved.

- Confidentiality: The electricity response messages should be confidential, i.e., if an adversary $\mathscr{A}$ captures the response messages, it cannot identify the encrypted messages.
- Authenticity and data integrity: BGs and HAN users should be authenticated by CC and BGs, respectively. Meanwhile, if an adversary $\mathscr{A}$ modifies the electricity demand, the malicious operations can be detected.
- Privacy preservation of electricity demand: The users' electricity demand should not be disclosed to the unauthorized/untrusted entities. Even if an adversary $\mathscr{A}$ hacks into the database of BGs and CC, it can also not identify the contents of ciphertexts.
- Forward secrecy of users' session keys: It should be ensured that the exposure of users' session keys corresponding to a given time period does not enable an adversary to decrypt any prior time period messages. Specifically, if an adversary $\mathscr{A}$ compromises a HAN user, $\mathscr{A}$ cannot get its previous electricity information. As a result, forward secrecy can be achieved.
- Evolution of users' private keys: The evolution of users' private keys should be achieved, i.e., if an adversary $\mathscr{A}$ compromises any previous private key of a HAN user, $\mathscr{A}$ cannot use it currently or in the future.

2.2.3   Design Goal

Our design goal is to develop an efficient PDR scheme with adaptive key evolution. In specific, the following three objectives should be achieved:

- The proposed scheme should achieve the privacy preservation of electricity demand, the demand's source authentication and data integrity, the confidentiality of the response messages.
- The proposed scheme should achieve efficient forward secrecy, i.e., evolution of users' session keys should be cost-effective in terms of computation and communication overhead.
- The proposed scheme should achieve adaptively key evolution, i.e., BG can adaptively control the frequency of key evolution by considering the balance of security level and communication efficiency.

2.3   Methodologies

In this section, we propose the PDR scheme, which consists of four phases: system initialization, demand aggregation, demand processing and response, and key evolution.

2.3.1   System Initialization

We assume a control center (CC) will bootstrap the whole system. Specifically, given the security parameter κ, CC firstly generates the bilinear parameters $(q, P, \mathbb{G}, \mathbb{G}_T, e)$ by running $\mathscr{G}en(\kappa)$, and chooses one secure symmetric encryption algorithm $Enc()$, e.g., AES, and three secure cryptographic hash functions H_1, H_2 and H_3, where $H_1, H_2 : \{0, 1\}^* \rightarrow \mathbb{G}$ and $H_3 : \mathbb{G}_T \rightarrow \mathbb{Z}_q^*$. In addition, CC also chooses a random number $\alpha \in \mathbb{Z}_q^*$, and computes $Q = \alpha P$ and $\mathscr{P}_{CC} = \alpha H_1(ID_{CC})$, where ID_{CC} is the identity string of CC. CC also calculates the homomorphic encryption's public key (N, g), and the corresponding private key (λ, μ). Finally, CC publishes the system parameters as pubs $= \{q, P, \mathbb{G}, \mathbb{G}_T, e, Q, H_1, H_2, H_3, N, g, Enc()\}$ and keeps the master key $(\lambda, \mu, \alpha, \mathscr{P}_{CC})$ secretly. When $BG_i (i = 1, 2, \cdots, n)$ registers itself into the system, CC runs the following steps:

- *Step-1:* CC computes the identity-based private key $SK_{BG_i} = \alpha H_1(ID_{CC} \| ID_{BG_i})$, where ID_{CC} and ID_{BG_i} are the identity strings of CC and BG_i, respectively.
- *Step-2:* CC grants SK_{BG_i} to BG_i through a secure channel [19].
   After receiving SK_{BG_i}, BG_i can non-interactively share a session key K_{BG_i-CC} with CC. BG_i computes $K_{BG_i-CC} = H_3(e(SK_{BG_i}, H_1(ID_{CC})))$, and CC computes $K_{BG_i-CC} = H_3(e(H_1(ID_{CC} \| ID_{BG_i}), \mathscr{P}_{CC}))$. The correctness is shown as follows:

$$K_{BG_i-CC} = H_3(e(SK_{BG_i}, H_1(ID_{CC})))$$
$$= H_3(e(\alpha H_1(ID_{CC}||ID_{BG_i}), H_1(ID_{CC})))$$
$$= H_3(e(H_1(ID_{CC}||ID_{BG_i}), H_1(ID_{CC}))^{\alpha}) \qquad (2.1)$$
$$= H_3(e(H_1(ID_{CC}||ID_{BG_i}), \alpha H_1(ID_{CC})))$$
$$= H_3(e(H_1(ID_{CC}||ID_{BG_i}), \mathscr{P}_{CC}))$$

Next, $BG_i (i = 1, 2, \cdots, n)$ chooses a random number $S_{BG_i} \in \mathbb{Z}_q^*$ as its master key and computes $Q_{BG_i} = S_{BG_i} P$ and its private point $\mathscr{P}_{BG_i} = S_{BG_i} H_1(ID_{BG_i})$. When a HAN user $U_{ij} (j = 1, 2, \cdots, m)$ registers itself into the BG_i, BG_i runs the following steps:

- *Step-1:* BG_i computes the identity-based private key $SK_{U_{ij}} = S_{BG_i} H_1(ID_{BG_i} || ID_{U_{ij}})$, where ID_{BG_i} and $ID_{U_{ij}}$ are the identity strings of BG_i and U_{ij}, respectively.
- *Step-2:* BG_i grants $SK_{U_{ij}}$ to U_{ij} through a secure channel [19].

After receiving $SK_{U_{ij}}$, U_{ij} can non-interactively share a session key $K_{U_{ij}-BG_i}$ with BG_i. U_{ij} computes $K_{U_{ij}-BG_i} = H_3(e(SK_{U_{ij}}, H_1(ID_{BG_i})))$, and BG_i computes $K_{U_{ij}-BG_i} = H_3(e(H_1(ID_{BG_i} || ID_{U_{ij}}), \mathscr{P}_{BG_i}))$. Similar to Eq. (2.1), the correctness is shown as follows:

$$K_{U_{ij}-BG_i} = H_3(e(SK_{U_{ij}}, H_1(ID_{BG_i})))$$
$$= H_3(e(S_{BG_i} H_1(ID_{BG_i} || ID_{U_{ij}}), H_1(ID_{BG_i})))$$
$$= H_3(e(H_1(ID_{BG_i} || ID_{U_{ij}}), H_1(ID_{BG_i}))^{S_{BG_i}}) \qquad (2.2)$$
$$= H_3(e(H_1(ID_{BG_i} || ID_{U_{ij}}), S_{BG_i} H_1(ID_{BG_i})))$$
$$= H_3(e(H_1(ID_{BG_i} || ID_{U_{ij}}), \mathscr{P}_{BG_i}))$$

2.3.2 Demand Aggregation

As shown in Fig. 2.2, each HAN user $U_{ij} \in BG_i (i = 1, 2, \cdots, n, j = 1, 2, \cdots, m)$ uses the smart meter to collect electricity demand d_{ij}, and performs the following steps:

- *Step-1:* U_{ij} chooses a random number $r_{ij} \in \mathbb{Z}_N^*$ and computes $C_{U_{ij}} = g^{d_{ij}} \cdot r_{ij}^N \mod N^2$.
- *Step-2:* U_{ij} uses the private key $SK_{U_{ij}}$ to make an identity-based signature $\sigma_{U_{ij}}$ on M, where $M = C_{U_{ij}} || ID_{BG_i} || ID_{U_{ij}} || TS$, TS is the current timestamp. Firstly, U_{ij} picks a random number $r_{ij} \in \mathbb{Z}_q^*$, and computes $U = r_{ij} P$, $V = SK_{U_{ij}} + r_{ij} H_2(M, U)$. The signature on M is the pair $\sigma_{U_{ij}} = <U, V>$.
- *Step-3:* U_{ij} sends the encrypted electricity demand $C_{U_{ij}} || ID_{BG_i} || ID_{U_{ij}} || TS || \sigma_{U_{ij}}$ to the BG_i.

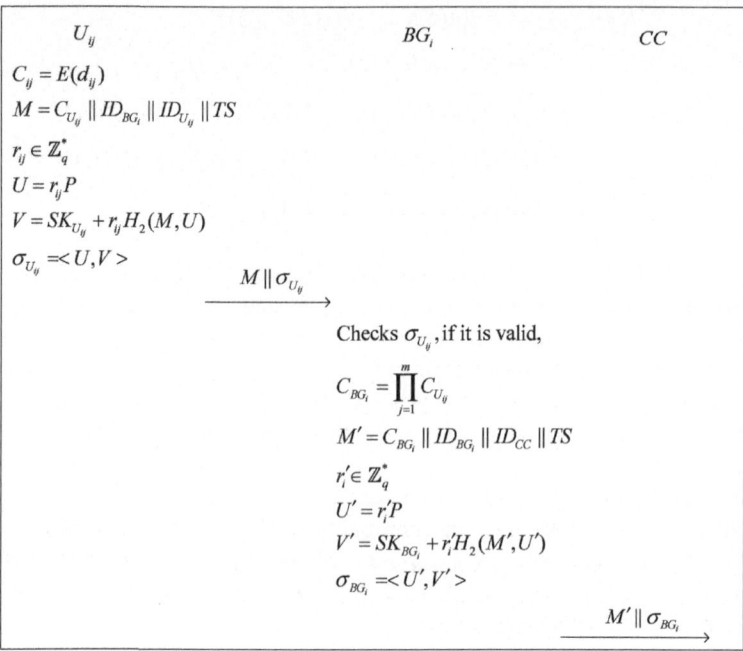

Fig. 2.2 Demand aggregation

After receiving encrypted electricity demand $C_{U_{ij}}||ID_{BG_i}||ID_{U_{ij}}||TS||\sigma_{U_{ij}}$, BG_i firstly checks signature $\sigma_{U_{ij}}$ to verify its validity. BG_i accepts the signature if the following equation holds.

$$e(P, V) = e(Q_{BG_i}, H_1(ID_{BG_i}||ID_{U_{ij}}))e(U, H_2(M, U)) \qquad (2.3)$$

where $M = C_{U_{ij}}||ID_{BG_i}||ID_{U_{ij}}||TS$. The correctness is shown as follows:

$$
\begin{aligned}
e(P, V) &= e(P, SK_{U_{ij}} + r_{ij}H_2(M, U)) \\
&= e(P, SK_{U_{ij}})e(P, r_{ij}H_2(M, U)) \\
&= e(P, S_{BG_i}H_1(ID_{BG_i}||ID_{U_{ij}}))e(r_{ij}P, H_2(M, U)) \qquad (2.4) \\
&= e(S_{BG_i}P, H_1(ID_{BG_i}||ID_{U_{ij}}))e(U, H_2(M, U)) \\
&= e(Q_{BG_i}, H_1(ID_{BG_i}||ID_{U_{ij}}))e(U, H_2(M, U))
\end{aligned}
$$

After the validity checking, BG_i performs the following steps based on multiple received $C_{U_{ij}}$ $(j = 1, \cdots, m)$:

- *Step-1:* BG_i computes the aggregated demand on $C_{U_{i1}}, C_{U_{i2}}, \cdots, C_{U_{im}}$ as $C_{BG_i} = \prod_{j=1}^{m} C_{U_{ij}}$.

- *Step-2:* BG_i uses the private key SK_{BG_i} to make an identity-based signature σ_{BG_i} on M', where $M' = C_{BG_i}||ID_{BG_i}||ID_{CC}||TS$, TS is the current timestamp. Firstly, BG_i picks a random number $r_i' \in \mathbb{Z}_q^*$, and computes $U' = r_i'P$ and $V' = SK_{BG_i} + r_i'H_2(M', U')$. The signature on M' is the pair $\sigma_{BG_i} = < U', V' >$. Then BG_i sends the aggregated result $C_{BG_i}||ID_{BG_i}||ID_{CC}||TS||\sigma_{BG_i}$ to the CC.

2.3.3 Demand Processing and Response

As shown in Fig. 2.3, upon receiving n encrypted electricity demand $C_{BG_i}||ID_{BG_i}||$ $ID_{CC}||TS||\sigma_{BG_i}(i = 1, 2, \cdots, n)$, CC firstly checks signature σ_{BG_i} to verify its validity. CC accepts the signature if the following equation holds

$$e(P, V') = e(Q, H_1(ID_{CC}||ID_{BG_i}))e(U', H_2(M', U')) \tag{2.5}$$

Where $M' = C_{BG_i}||ID_{BG_i}||ID_{CC}||TS$. The correctness is shown as follows:

$$\begin{aligned}
e(P, V') &= e(P, SK_{BG_i} + r_i'H_2(M', U')) \\
&= e(P, SK_{BG_i})e(P, r_i'H_2(M', U')) \\
&= e(P, \alpha H_1(ID_{CC}||ID_{BG_i}))e(r_i'P, H_2(M', U')) \\
&= e(\alpha P, H_1(ID_{CC}||ID_{BG_i}))e(U', H_2(M', U')) \\
&= e(Q, H_1(ID_{CC}||ID_{BG_i}))e(U', H_2(M', U'))
\end{aligned} \tag{2.6}$$

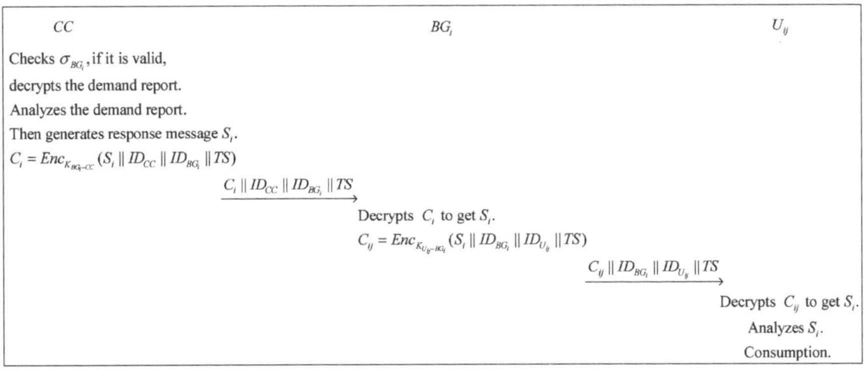

Fig. 2.3 Demand processing and response

After the validity checking, CC performs the following steps to read the aggregated demand C, where C is implicitly formed by

$$C = \prod_{i=1}^{n} C_{BG_i}$$

$$= \prod_{i=1}^{n} (\prod_{j=1}^{m} C_{U_{ij}})$$

$$= \prod_{i=1}^{n} (\prod_{j=1}^{m} g^{d_{ij}} \cdot r_{ij}^{N} \bmod N^2) \qquad (2.7)$$

$$= \prod_{i=1}^{n} (g^{\sum_{j=1}^{m} d_{ij}} \cdot (\prod_{j=1}^{m} r_{ij})^N \bmod N^2)$$

$$= g^{\sum_{i=1}^{n} (\sum_{j=1}^{m} d_{ij})} \cdot (\prod_{i=1}^{n} (\prod_{j=1}^{m} r_{ij}))^N \bmod N^2$$

- *Step-1:* By taking $M = \sum_{i=1}^{n} (\sum_{j=1}^{m} d_{ij})$ and $R = \prod_{i=1}^{n} (\prod_{j=1}^{m} r_{ij})$, the report $C = g^M \cdot R^N \bmod N^2$ is still a ciphertext of Paillier cryptosystem. Therefore, CC can use the master key (λ, μ) to recover M as $M = \sum_{i=1}^{n} (\sum_{j=1}^{m} d_{ij})$. Similarly, CC can recover BG_i's aggregated electricity demand as $\sum_{j=1}^{m} d_{ij}$.
- *Step-2:* After analyzing the real-time electricity demand $\sum_{i=1}^{n} (\sum_{j=1}^{m} d_{ij})$ and $\sum_{j=1}^{m} d_{ij} (i = 1, 2, \cdots, n)$, CC generates the response message $S_i (0 < S_i \leq 1)$ for $BG_i (i = 1, 2, \cdots, n)$, respectively [9], where S_i is a scale coefficient. For example, the electricity demand from BG_i is $\sum_{j=1}^{m} d_{ij}$=20,000 kW/h, however, CC would like to provide 16,000 kW/h considering the electricity generation and the total electricity demand $\sum_{i=1}^{n} (\sum_{j=1}^{m} d_{ij})$. Then CC sets $S_i = 0.8$. If electricity consumption from BG_i is more than 16,000 kW/h, the electricity tariff will be higher than before.
- *Step-3:* CC sends $C_i || ID_{CC} || ID_{BG_i} || TS$ to $BG_i (i = 1, 2, \cdots, n)$, respectively, where $C_i = Enc_{K_{BG_i - CC}} (S_i || ID_{CC} || ID_{BG_i} || TS)$ and TS is the current timestamp.
- *Step-4:* Upon receiving $C_i || ID_{CC} || ID_{BG_i} || TS$, BG_i decrypts C_i to get S_i. Then BG_i forwards $C_{ij} || ID_{BG_i} || ID_{U_{ij}} || TS$ to the HAN user U_{ij}, where $C_{ij} = Enc_{K_{U_{ij} - BG_i}} (S_i ||$
$ID_{BG_i} || ID_{U_{ij}} || TS)$ and TS is the current timestamp.
- *Step-5:* After receiving $C_{ij} || ID_{BG_i} || ID_{U_{ij}} || TS$, HAN user U_{ij} decrypts C_{ij} to get S_i. Then U_{ij} analyzes S_i and determines to shift power use from peak times to non-peak times for lower electricity bills [9].

2.3.4 Key Evolution

Firstly, we extend the identity string ID to $ID || d$, where d represents the expiry date. Note that the extension does not influence the previous PDR scheme. For the

HAN user U_{ij}, the identity string $ID_{U_{ij}}||d$ is only valid before the specified expiry date d. After d, the corresponding private key $SK_{U_{ij}||d}$ is automatically revoked if a new private key is not generated by BG_i. If the unit of d is chosen as 1 day [20], the lifetime of each private key is also the same. As shown in Fig. 2.4, the proposed key evolution mechanism is comprised of many rounds. At the end of $Round_i (i = 1, 2, \cdots)$, $Round_{i+1}$'s keys will be generated by key evolution algorithm as described in Fig. 2.5. The time interval of each round can be calculated by the number of its keys. For instance, if the number of $Round_{i+1}$'s keys is δ, then the time interval of $Round_{i+1}$ is δ days since the lifetime of each key is 1 day. Next, we discuss the key evolution algorithm in detail. Specifically, at the end of $Round_i (i = 1, 2, \cdots)$, a HAN user U_{ij} can generate $Round_{i+1}$'s keys by performing the following steps:

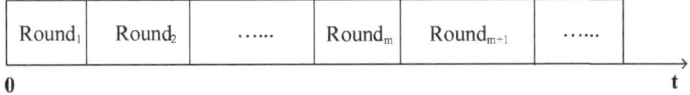

Fig. 2.4 Proposed key evolution mechanism

- *Step-1:* According to the security and efficiency requirement, U_{ij} chooses an integer l_{ij} as the number of evolving private keys. For example, when the security level is more important than efficiency, U_{ij} can choose a small integer for l_{ij}, e.g., $l_{ij} = 5$. On the other hand, when the efficiency is more important than security level, U_{ij} can choose a bigger integer for l_{ij}, e.g., $l_{ij} = 30$. we will further show the results in Sect. 2.5.
- *Step-2:* U_{ij} sends $C_{ij}||ID_{BG_i}||ID_{U_{ij}}||TS$ to BG_i, where $C_{ij} = Enc_{K_{U_{ij}-BG_i}}(l_{ij}|| ID_{BG_i}||ID_{U_{ij}}||TS)$ and TS is the current timestamp.
- *Step-3:* After receiving $C_{ij}||ID_{BG_i}||ID_{U_{ij}}||TS$, BG_i firstly decrypts C_{ij} to get l_{ij} with the session key $K_{U_{ij}-BG_i}$. Then BG_i checks whether $l_{ij} < m_{ij}$, where m_{ij} represents the maximum number of private keys that BG_i can assign to U_{ij}. If it does hold, BG_i does nothing. Otherwise, BG_i lets $l_{ij} = m_{ij}$. And then BG_i generates l_{ij} private keys as

$$SK_{U_{ij}||(d+1)} = S_{BG_i} H_1(ID_{BG_i}||ID_{U_{ij}}||(d+1))$$

$$SK_{U_{ij}||(d+2)} = S_{BG_i} H_1(ID_{BG_i}||ID_{U_{ij}}||(d+2))$$

$$\cdots \tag{2.8}$$

$$SK_{U_{ij}||(d+l_{ij})} = S_{BG_i} H_1(ID_{BG_i}||ID_{U_{ij}}||(d+l_{ij}))$$

The set of private keys is as follows:

$$\omega = \{SK_{U_{ij}||(d+1)}, SK_{U_{ij}||(d+2)}, \cdots, SK_{U_{ij}||(d+l_{ij})}\} \tag{2.9}$$

- *Step-4:* BG_i sends $C'_{ij} \| ID_{BG_i} \| ID_{U_{ij}} \| TS$ to U_{ij}, where $C'_{ij} = Enc_{K_{U_{ij}-BG_i}}(\omega \| ID_{BG_i} \| ID_{U_{ij}} \| TS)$ and TS is the current timestamp.
- *Step-5:* After receiving $C'_{ij} \| ID_{BG_i} \| ID_{U_{ij}} \| TS$ to U_{ij}, U_{ij} firstly recovers ω with the session key $K_{U_{ij}-BG_i}$. Then U_{ij} deletes previous session and processing information and stores ω secretly. Thus, the secure private key evolution is achieved.
- *Step-6:* On the date $d + a (a = 1, \cdots, l_{ij})$, U_{ij} firstly deletes the previous private key. Then U_{ij} can non-interactively share a new session key $K'_{U_{ij}-BG_i}$ with BG_i. U_{ij} computes

$$K'_{U_{ij}-BG_i} = H_3(e(SK_{U_{ij}\|(d+a)}, H_1(ID_{BG_i}))) \tag{2.10}$$

And BG_i computes

$$K'_{U_{ij}-BG_i} = H_3(e(H_1(ID_{BG_i} \| ID_{U_{ij}} \|(d + a)), \mathscr{P}_{BG_i})) \tag{2.11}$$

The correctness is shown as follows:

$$
\begin{aligned}
K'_{U_{ij}-BG_i} &= H_3(e(SK_{U_{ij}\|(d+a)}, H_1(ID_{BG_i}))) \\
&= H_3(e(H_1(ID_{BG_i} \| ID_{U_{ij}} \|(d + a)), H_1(ID_{BG_i}))^{S_{BG_i}}) \\
&= H_3(e(H_1(ID_{BG_i} \| ID_{U_{ij}} \|(d + a)), \mathscr{P}_{BG_i}))
\end{aligned}
\tag{2.12}
$$

After that, U_{ij} deletes the previous session key $K_{U_{ij}-BG_i}$. Thus, even if an adversary $\mathscr{A}$ compromises U_{ij}, it cannot get any previous session key. Therefore, the forward secrecy of the session key is achieved.

2.4 Security Analysis

In this section, we analyze the security properties of the proposed PDR scheme in terms of authentication and data integrity of electricity demand, and the confidentiality of response messages, the privacy preservation of electricity demand, the forward secrecy of users' session keys, and the evolution of users' private keys.

2.4.1 Authenticity, Data Integrity and Confidentiality

As mentioned in Sect. 2.3.2, each HAN user's electricity demand and the aggregated demand are signed by the identity-based signature [21]. Since the signature is provably secure in the random oracle model [21], the source authentication and data integrity of electricity demand can be guaranteed. In addition, note that the

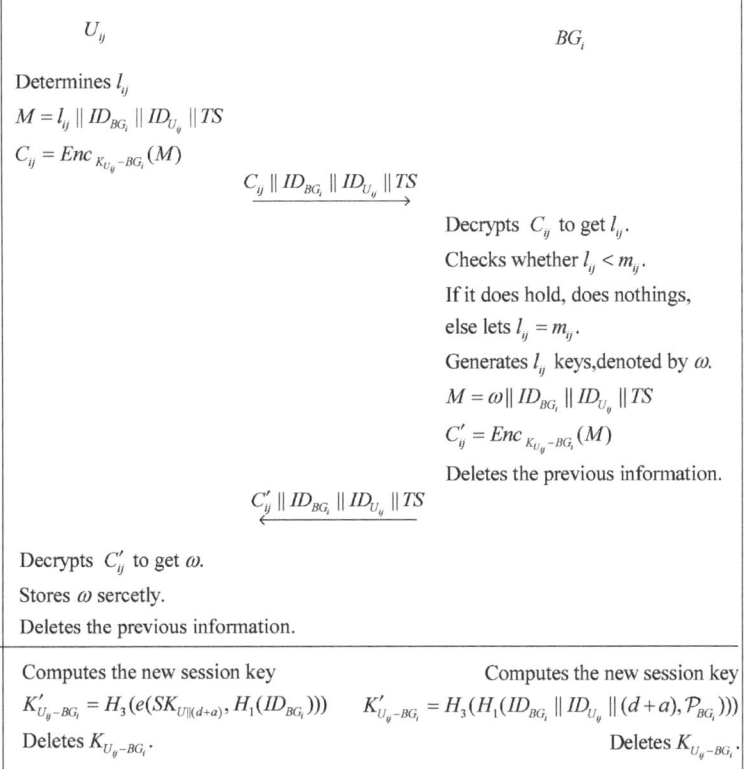

Fig. 2.5 Key evolution algorithm

HAN user U_{ij}'s private key $SK_{U_{ij}} = S_{BG_i} H_1(ID_{BG_i} || ID_{U_{ij}})$, namely $SK_{U_{ij}}$ is bound to BG_i. Thus, BG_i can identify whether a HAN user belongs to its administration domain. As mentioned in Eq. (2.3), a HAN user cannot pass the signature verification if it is not in BG_i's administration domain. On the other hand, in the proposed PDR scheme, when CC sends the response messages to $BG_i(i = 1, 2, \cdots, n)$, CC encrypts them as $C_i = Enc_{K_{BG_i - CC}}(S_i || ID_{CC} || ID_{BG_i} || TS)$. Then, BG_i forwards the response messages to $U_{ij}(j = 1, 2, \cdots, m)$ in the form of $C_{ij} = Enc_{K_{U_{ij} - BG_i}}(S_i || ID_{BG_i} || ID_{U_{ij}} || TS)$. Since C_i and C_{ij} are encrypted by AES [22], the confidentiality of response messages can be guaranteed.

2.4.2 Privacy Preservation of Electricity Demand

Since HAN user's electricity demand is a homomorphic encryption ciphertext [18], an adversary $\mathscr{A}$ cannot identify the corresponding electricity demand even though $\mathscr{A}$ eavesdrops the ciphertext. Moreover, since BG_i only aggregates and does not decrypt the electricity demands, $\mathscr{A}$ cannot get the electricity demand even if $\mathscr{A}$ compromises the BG_i's database. Finally, CC recovers the aggregated demands $\sum_{i=1}^{n}(\sum_{j=1}^{m} d_{ij})$ and $\sum_{j=1}^{m} d_{ij} (i = 1, 2, \cdots, n)$. However, since $\sum_{i=1}^{n}(\sum_{j=1}^{m} d_{ij})$ and $\sum_{j=1}^{m} d_{ij} (i = 1, 2, \cdots, n)$ are all aggregated results, even if $\mathscr{A}$ intrudes the CC's database, $\mathscr{A}$ still cannot get each HAN user's electricity demand. Therefore, the proposed PDR scheme preserves the electricity demand privacy.

2.4.3 Forward Secrecy of Users' Session Keys

The confidentiality of communication between HAN user and BG is achieved based on the secure session key. In the key evolution phase, after computing the new session key $K'_{U_{ij}-BG_i}$, U_{ij} deletes the previous session key $K_{U_{ij}-BG_i}$. As a result, even if $\mathscr{A}$ compromises the HAN user's U_{ij}, $\mathscr{A}$ cannot get any previous session key. Moreover, $\mathscr{A}$ cannot compute any previous session key as mentioned in Eq. (2.10) since the corresponding private key has been deleted. Therefore, the forward secrecy of users' session keys is achieved in the proposed PDR scheme.

Furthermore, we analyze the information leakage in both schemes without forward secrecy (denoted by NFS) and with forward secrecy (denoted by FS). The information leakage happens when the encrypted messages are decrypted by an unauthorized adversary [23]. The number of information leakages can be calculated by the time interval in which the encrypted messages cannot be guaranteed to be confidential. As shown in Fig. 2.6, t_0 and t_1 represent the times when U_{ij} is compromised and when the system detects the attack and revokes U_{ij}, respectively. T_c represents the time interval between $t = 0$ and $t = t_0$. X_i and T_d represent the time interval of $Round_i$ and the time delay of system detection, respectively. In probability theory and statistics, the Poisson distribution is a discrete probability distribution that expresses the probability of a given number of events occurring in a fixed interval of time and/or space if these events occur with an average rate and independently of the time since the last event [24]. Note that X_i and T_d can be seen as the discrete events in the visual spaces "the time interval of $Round_i$" and "the time delay of system detection", respectively. And "the time interval of $Round_i$" and "the time delay of system detection" both occur with an average rate and independently of the time since the last event. Therefore, we can model that X_i and T_d follow the Poisson distribution with intensity λ_x and λ_d, respectively, then $P(X_i = k) = \frac{e^{-\lambda_x}\lambda_x^k}{k!}$ and $P(T_d = k) = \frac{e^{-\lambda_d}\lambda_d^k}{k!}$. Next, we discuss the number of information leakage in both NFS and FS. For NFS, all messages encrypted before $t = t_1$ can be decrypted since the session key has not been evolved since $t = 0$. Thus, the number of information leakage for NFS is $T_c + \lambda_d$. In comparison, for FS, since a user U_{ij} only stores the current round's keys and has deleted the previous rounds' keys, the previous rounds' messages cannot be decrypted even if U_{ij} is compromised. Therefore, in the worst case of FS when t_0 is at the end of $Round_i (i = 1, 2, \cdots)$, an adversary can decrypt the messages encrypted in both $Round_i$ and the following λ_d time interval. Thus, the number of information leakage is $\lambda_x + \lambda_d (t_0 \geq X_1)$. And in the best case of FS when t_0 is at the beginning of a round, an adversary only can decrypt the messages encrypted in the following λ_d time interval. Thus, the number of information leakages is λ_d.

The comparison of the number of information leakages between FS and NFS is shown in Fig. 2.7. It can be seen that FS significantly reduces the number of information leakages compared with NFS. In addition, it is observed that when λ_d is constant, the number of information leakages increases with the increased λ_x. For example, when $\lambda_d = 5$ and $\lambda_x = 10$, as shown in Fig. 2.7a, the number of information leakages is in the interval [5,15]. However, when $\lambda_d = 5$ and $\lambda_x = 20$, as shown in Fig. 2.7c, the number of information leakages is in the interval [5,25]. Note that λ_x represents the intensity of X_i, where X_i is the time interval of $Round_i$. As mentioned in Sect. 2.3.4, X_i can be controlled by m_{ij} since m_{ij} is the upper bound of X_i. When m_{ij} is set small enough, X_i is also small. Further λ_x is also

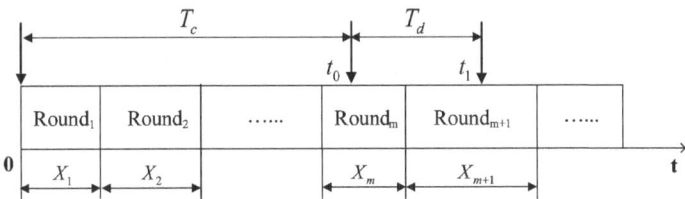

t_0 : the compromised time of U_{ij}

t_1 : the revocation time of U_{ij}

T_c : the time interval between $t = 0$ and $t = t_0$

T_d : the time delay of system detection

$X_i (i = 1, 2, \cdots)$: the time interval of $Round_i$

Fig. 2.6 Information leakage model

small since λ_x is the average value of $X_i (i = 1, 2, \cdots)$ [24]. Thus, the small λ_x enables the number of information leakage to be low. In comparison, when m_{ij} is set bigger, X_i and λ_x can be relatively increased, thereby make the number of information leakage high. Therefore, in our key evolution technique, when m_{ij} is set relatively bigger when the evolution of session keys and private keys is sparse, the number of information leakage is higher. As a result, the security level is degraded. On the other hand, when m_{ij} is set smaller when the evolution of session keys and private keys is frequent, the number of information leakages can be decreased and the security level can be upgraded. However, the smaller m_{ij} will lead to heavy communication overhead. We will discuss that in Sect. 2.5.1.

2.4.4   Evolution of Users' Private Keys

In the key evolution phase, U_{ij} firstly sends l_{ij} to BG_i by the symmetric encryption algorithm. Then according to l_{ij} provided by U_{ij}, BG_i generates the set of private keys ω and further sends ω to U_{ij} by the symmetric encryption algorithm. Thus, even if an adversary $\mathscr{A}$ eavesdrops on the communication between BG_i and U_{ij}, it cannot get any information about ω. On the other hand, even if an adversary $\mathscr{A}$ compromises any previous private key, it cannot deduce current or future private keys since the discrete logarithm problem ensures the private keys' security [19]. Therefore, the secure private key evolution of a user is achieved in the proposed PDR scheme.

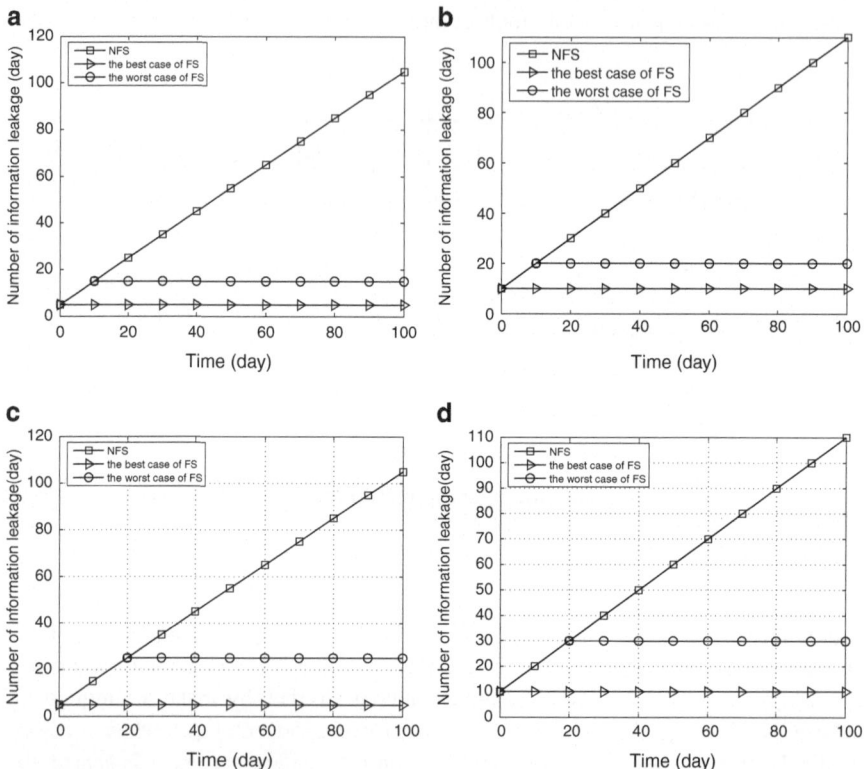

Fig. 2.7 Comparison of the number of information leakage. (a) $\lambda_x=10$ and $\lambda_d=5$. (b) $\lambda_x=10$ and $\lambda_d=10$. (c) $\lambda_x=20$ and $\lambda_d=5$. (d) $\lambda_x=20$ and $\lambda_d=10$

Finally, we present the comparison results of security levels in Table 2.1. It can be seen that the scheme [16] only achieves the confidentiality, the scheme [25] achieves confidentiality and data integrity, the scheme [3] achieves confidentiality,

Table 2.1 Comparison of
security level

	[16]	[25]	[3]	[17]	PDR
Confidentiality	✓	✓	✓	✓	✓
Authenticity			✓	✓	✓
Data integrity		✓	✓	✓	✓
Forward secrecy				✓	✓
Private key evolution					✓

data integrity and authenticity, and the scheme [17] achieves confidentiality, data integrity, authenticity, and forward secrecy. The proposed PDR scheme achieves additional private key evolution compared with the scheme [17].

2.5   Performance Evaluation

In this section, we evaluate the communication and computation overheads of the session key evolution between user U_{ij} and BG_i in both the scheme [17] and PDR.

2.5.1   Communication Overhead

In PDR, for evolving l_{ij} session keys, U_{ij} firstly sends a message to BG_i. The message is in the form of $C_{U_{ij}}||ID_{BG_i}||ID_{U_{ij}}||TS$, where $C_{U_{ij}} = Enc_{K_{U_{ij}-BG_i}}(l_{ij}||ID_{BG_i}||ID_{U_{ij}}||TS)$. If we choose AES ciphertext with 256-bit and set $|BG_i| + |U_{ij}| + |TS|$ as 80-bit length, the whole message size is $256 + 80 = 336$ bits. Then, BG_i checks if $l_{ij} < m_{ij}$, if it does hold, BG_i does nothing, else lets $l_{ij} = m_{ij}$. And then BG_i responds to a message in the form of $C'_{ij}||ID_{BG_i}||ID_{U_{ij}}||TS$ to U_{ij}, where $C'_{ij} = Enc_{K_{U_{ij}-BG_i}}(\omega||ID_{BG_i}||ID_{U_{ij}}||TS)$, $\omega = \{SK_{U_{ij}||(d+1)}, SK_{U_{ij}||(d+2)}, \cdots, SK_{U_{ij}||(d+l_{ij})}\}$. If we choose $\mathbb{G}$ with 160-bit order and use point compression technique [26], the element in $\mathbb{G}$ is roughly 161-bit. If we choose AES ciphertext with 256-bit, C'_{ij} should be generated based on the 256-bit block encryption [22]. Thus, the size of C'_{ij} is $\lceil (161 * Min(l_{ij}, m_{ij}) + 80)/256 \rceil * 256$ bits, where $Min(l_{ij}, m_{ij})$ is the minimum between l_{ij} and m_{ij}. Note that for evolving l_{ij} session keys, the whole evolving process will be run $\lceil l_{ij}/m_{ij} \rceil$ times. So the total communication overhead is $\lceil l_{ij}/m_{ij} \rceil (\lceil (161 * Min(l_{ij}, m_{ij}) + 80)/256 \rceil * 256 + 80 + 336)$ bits. In comparison, in the scheme [17], for evolving each session key, HAN user i firstly sends an RSA ciphertext to BG j in the form of $\{i||j||g^a\}_{\{encr\}PubBAN\_GW_j}$. Then BAN GW j responds to an RSA ciphertext to HAN user i in the form of $\{i||j||g^a||g^b\}_{\{encr\}PubHAN\_GW_i}$. Finally, HAN user i sends an AES ciphertext to BG j in the form of $\{M_i||T_i||HMAC_{K_i}\}_{\{encr\}K_i}$. The overall communication overhead consists of two RSA ciphertexts and one AES ciphertext. Thus, the overall communication overhead is $2 * 1024 + 256 = 2304$ bits if we choose 1024-bit RSA and 256-bit AES. Therefore, the total communication overhead for evolving l_{ij} session keys is $2304 * l_{ij}$ bits.

The communication overhead for different numbers of evolving session keys is shown in Fig. 2.8. When the number of evolving session keys is small, the communication overhead is low in both PDR and the scheme [17]. Then the communication overhead increases with the increased number of keys. However, it should be noted that the increase is much faster in the case of the scheme [17]. PDR significantly reduces the communication overhead for the session key evolution. On the other hand, among three variants of PDR with different m_{ij}, the communication overhead decreases with the increased m_{ij} since the bigger m_{ij} reduces the frequency of key evolution. However, as mentioned in Sect. 2.4.3, when m_{ij} is increased, the number of information leakage is also increased and the security level is degraded. In our proposed PDR scheme, as mentioned in Sect. 2.3.4,

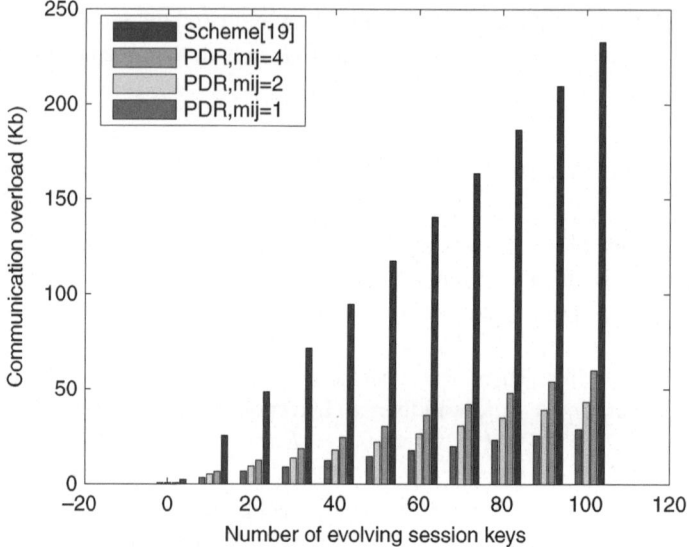

Fig. 2.8 Comparison of communication overhead

BG_i can adaptively adjust m_{ij} to balance the trade-off between the communication efficiency and security level.

2.5.2 Computation Overhead

Compared with pairing operations, RSA encryption/decryption and exponentiation operations in $\mathbb{G}$, the computation overhead of AES encryption/decryption and hash operations are negligible [27]. In the scheme [17], for evolving each session key, it requires 1 RSA encryption for HAN user i to generate the request packet. After receiving the ciphertext from HAN user i, BG j decrypts the request packet including 1 RSA decryption and computes the new session key with 1 exponentiation operation in $\mathbb{Z}_q^*$. In addition, BG j sends an encrypted response message including 1 RSA encryption. Then HAN user i decrypts the response message with 1 RSA decryption and computes the new session key including 1 exponentiation operation in $\mathbb{Z}_q^*$. Denote the computation overhead of an exponentiation operation in $\mathbb{Z}_q^*$, an RSA encryption and an RSA decryption by C_e, RSA_e and RSA_d, respectively. Thus, the total computation overhead for evolving a session key is $2 * (C_e + RSA_e + RSA_d)$. Therefore, the total computation overhead for evolving l_{ij} session keys is $2 * l_{ij} * (C_e + RSA_e + RSA_d)$.

In comparison, in PDR, as described in Sect. 2.3.4, BG_i computes l_{ij} new private keys and l_{ij} new session keys with l_{ij} multiplication operations in $\mathbb{G}$ and l_{ij} pairing operations, respectively. U_{ij} computes the new session keys with l_{ij} pairing operations. Note that the above computation overhead is constant even if m_{ij} varies. Therefore, in this section, we do not consider the variants of PDR with different m_{ij}. We will discuss the balance between the communication efficiency and security level in Sect. 2.5.1. Denote the computation overhead of a multiplication operation in $\mathbb{G}$ and a pairing operation by C_m and C_p, respectively. Thus, the total computation overhead is $l_{ij} * C_m + 2l_{ij} * C_p$.

We conducted an experiment on a 3.0 GHz-processor, 1 GB memory computing machine with MIRACL [28] and Pbc [29] libraries to study the execution time. For $\mathbb{G}$ over the FST curve, a single multiplication operation costs 1.1 ms and the corresponding pairing operation costs 3.1 ms. Meantime, a 1024-RSA decryption and a 1024-RSA encryption cost 3.88 ms and 0.02 ms, respectively. An exponentiation operation in $\mathbb{Z}_q^*(|q| = 1024)$ costs 0.64 ms. The comparison of computation overhead is shown in Fig. 2.9. We can see that PDR achieves lower execution times compared with the scheme [17].

2.6 Related Works

Forward secrecy is a property that ensures that the messages of prior time period are confidential even if the current time period's key has been compromised [15]. Kate et al. [30] present an improved forward secrecy scheme for onion routing anonymity networks, its computation and communication overheads are significantly less than the previous schemes. Chen et al. [31] propose an efficient approach to establish security links in wireless sensor networks. The proposed scheme only requires small memory size and can achieve forward secrecy. Forward secrecy can be implemented by the key evolution technique, which generates the new keys based on the old ones. Liu et al. [32] propose a key evolution technique for sensor networks. The technique can ensure forward secrecy and achieve viable trade-offs between security and resource consumption. Libert et al. [20] propose the key evolution systems in untrusted update environments. The systems implement an efficient generic

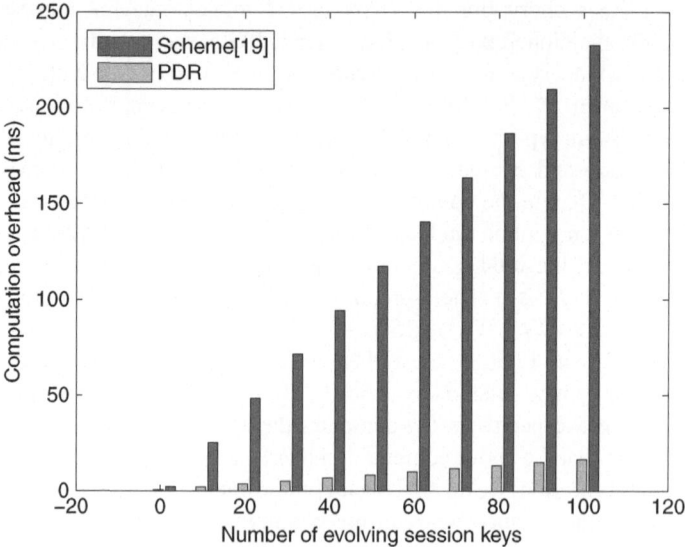

Fig. 2.9 Comparison of computation overhead

construction and can be extended a forward-secure public key encryption scheme, although the importance of forward secrecy and key evolution, how to design a data aggregation scheme with forward secrecy of session keys, and the evolution of users' private keys in smart grids is a challenging issue. Fouda et al. [17] propose a lightweight message authentication scheme achieving forward secrecy. Specifically, in the proposed scheme, HAN users can achieve mutual authentication with BG. Detailed security analysis shows that the proposed scheme can satisfy confidentiality, data integrity, authenticity and forward secrecy. Since the scheme adopts RSA public key signature algorithm and Diffie-Hellman exchange protocol to evolve the session keys between HAN users and BG, the computation and communication overhead are heavy.

Homomorphic encryption achieves certain algebraic operations on the plaintext to be performed directly on the ciphertext and has been used in many data aggregation schemes [33,34]. These schemes are very promising and have triggered considerable research work [3, 16, 25] in smart grids as follows. Li et al. [16] present a distributed incremental data aggregation approach. To protect user privacy, homomorphic encryption is used to secure the data en route. Seo et al. [25] propose a secure and efficient power management mechanism leveraging a homomorphic data aggregation and capability-based power distribution. The proposed mechanism enables to gather the power demands of users securely. Lu et al. [3] propose a privacy-preserving aggregation scheme for secure and efficient smart grid communications. It realizes the multi-dimensional data aggregation approach based on the homomorphic Paillier cryptosystem. These schemes assume that the session keys between HAN users and building area network gateway (BG) are unchanged. However, once an adversary $\mathscr{A}$ compromises the session keys, $\mathscr{A}$ can decrypt any previous response message.

2.7 Summary

In this chapter, we have proposed an efficient PDR scheme with adaptive key evolution. It realizes secure and efficient electricity demand aggregation and response based on homomorphic encryption and key evolution techniques. Security analysis has demonstrated that PDR can achieve privacy preservation of electricity demand, forward secrecy of users' session keys, and evolution of users' private keys. Performance evaluation further demonstrates its efficiency in terms of computation and communication overhead. In addition, PDR can adaptively control the key evolution to balance the trade-off between the communication efficiency and security level.

References

1. R. Deng, J. Chen, X. Cao, Y. Zhang, S. Maharjan, and S. Gjessing, "Sensing-performance tradeoff in cognitive radio enabled smart grid," *IEEE Transactions on Smart Grid*, vol. 4, no. 1, pp. 302–310, 2013.
2. H. Liang, B. Choi, W. Zhuang, and X. Shen, "Towards optimal energy store-carry-and-deliver for phevs via v2g system," in *IEEE INFOCOM*, 2012, pp. 1674–1682.
3. R. Lu, X. Liang, X. Li, X. Lin, and X. Shen, "Eppa: An efficient and privacy preserving aggregation scheme for secure smart grid communications," *IEEE Transactions on Parallel and Distributed Systems*, vol. 23, no. 9, pp. 1621–1631, 2012.
4. H. Li, X. Lin, H. Yang, X. Liang, R. Lu, and X. Shen, "Eppdr: An efficient privacy-preserving demand response scheme with adaptive key evolution in smart grid," *IEEE Transactions on Parallel and Distributed Systems*, http://www.computer.org/csdl/trans/td/preprint/06506075-abs.html.
5. H. Liang, B. Choi, A. Abdrabou, W. Zhuang, and X. Shen, "Decentralized economic dispatch in microgrids via heterogeneous wireless networks," *IEEE Journal on Selected Areas in Communications*, vol. 30, no. 6, pp. 1061–1074, 2012.
6. M. Wen, R. Lu, J. Lei, H. Li, X. Liang, and X. Shen, "Ecq: An efficient conjunctive query scheme over encrypted multidimensional data in smart grid," in *IEEE GLOBECOM*, Atlanta, GA, USA, 2013.
7. X. Lu, W. Wang, and J. Ma, "An empirical study of communication infrastructures towards the smart grid: Design, implementation, and evaluation," *IEEE Transactions on Smart Grid*, vol. 4, no. 1, pp. 170–183, 2013.
8. M. Wen, R. Lu, J. Lei, H. Li, X. Liang, and X. Shen, "Sesa: An efficient searchable encryption scheme for auction in emerging smart grid marketing," *Security and Communication Networks*, http://onlinelibrary.wiley.com/doi/10.1002/sec.699/full.
9. F. Rahimi and A. Ipakchi, "Demand response as a market resource under the smart grid paradigm," *IEEE Transactions on Smart Grid*, vol. 1, no. 1, pp. 82–88, 2010.

10. X. Li, X. Liang, R. Lu, H. Zhu, X. Lin, and X. Shen, "Securing smart grid: Cyber attacks, countermeasures and challenges," *IEEE Communications Magazine*, vol. 58, no. 8, pp. 38–45, 2012.

11. X. Liang, X. Li, R. Lu, X. Lin, and X. Shen, "Udp: Usage-based dynamic pricing with privacy preservation for smart grid," *IEEE Transactions on Smart Grid*, vol. 4, no. 1, pp. 141–150, 2013.

12. H.Li, R.Lu, L.Zhou, B.Yang, and X.Shen, "An efficient merkle tree based authentication scheme for smart grid," *IEEE Systems Journal*, http://ieeexplore.ieee.org/xpls/abs_all.jsp? arnumber=6563123.

13. H. Li, X. Liang, R. Lu, X. Lin, and X. Shen, "Edr: An efficient demand response scheme for achieving forward secrecy in smart grid," in *IEEE GLOBECOM*, 2012, pp. 929–934.

14. J. Xia and Y. Wang, "Secure key distribution for the smart grid," *IEEE Transactions on Smart Grid*, vol. 3, no. 3, pp. 1437–1443, 2012.

15. R. Canetti, S. Halevi, and J. Katz, "A forward-secure public-key encryption scheme," *Advances in Eurocrypt 2003*, pp. 646–646, 2003.

16. F. Li, B. Luo, and P. Liu, "Secure information aggregation for smart grids using homomorphic encryption," in *2010 IEEE International Conference on Smart Grid Communications (Smart-GridComm)*, 2010, pp. 327–332.

17. M. Fouda, Z. Fadlullah, N. Kato, R. Lu, and X. Shen, "A lightweight message authentication scheme for smart grid communications," *IEEE Transactions on Smart Grid*, vol. 2, no. 4, pp. 675–685, 2011.

18. P. Paillier, "Public-key cryptosystems based on composite degree residuosity classes," in *EUROCRYPT*, 1999, pp. 223–238.

19. D. Boneh and M. K. Franklin, "Identity-based encryption from the weil pairing," in *CRYPTO*, 2001, pp. 213–229.

20. B. Libert, J. Quisquater, and M. Yung, "Key evolution systems in untrusted update environments," *ACM Transactions on Information and System Security*, vol. 13, no. 4, p. 37, 2010.

21. B. Libert and J. Quisquater, "The exact security of an identity based signature and its applications," *Preprint available at http://eprint.iacr.org/2004/102*, 2004.

22. D. Stinson, *Cryptography: theory and practice*. CRC press, 2006.

23. J. Kelsey, "Compression and information leakage of plaintext," in *Fast Software Encryption 2002*. Springer, 2002, pp. 95–102.

24. S. Katti and A. Rao, "Handbook of the poisson distribution," *Technometrics*, vol. 10, no. 2, pp. 412–412, 1968.

25. D. Seo, H. Lee, and A. Perrig, "Secure and efficient capability-based power management in the smart grid," in *IEEE International Symposium on Parallel and Distributed Processing with Applications Workshops (ISPAW)*, 2011, pp. 119–126.

26. I. Blake, G. Seroussi, and N. Smart, "Pairings," *Advances in elliptic curve cryptography*, pp. 183–213, chapter 9, Cambridge University Press, 2005.

27. W. Dai, "Crypto++ 5.6.0 benchmarks," *http://www.cryptopp.com/benchmarks.html*, 2009.

28. "Miracl crypto," *https://certivox.com/solutions/miracl-crypto-sdk/*.

29. B. Lynn, "Pbc library," *http://crypto.stanford.edu/pbc/*.

30. A. Kate, G. Zaverucha, and I. Goldberg, "Pairing-based onion routing with improved forward secrecy," *ACM Transactions on Information and System Security*, vol. 13, no. 4, p. 29, 2010.

31. C. Chen, S. Huang, and I. Lin, "Providing perfect forward secrecy for location-aware wireless sensor networks," *EURASIP Journal on Wireless Communications and Networking*, vol. 2012, no. 1, p. 241, 2012.

32. Z. Liu, J. Ma, Q. Pei, L. Pang, and Y. Park, "Key infection, secrecy transfer, and key evolution for sensor networks," *IEEE Transactions on Wireless Communications*, vol. 9, no. 8, pp. 2643–2653, 2010.

33. J. Shi, R. Zhang, Y. Liu, and Y. Zhang, "Prisense: privacy-preserving data aggregation in people-centric urban sensing systems," in *INFOCOM*, 2010, pp. 1–9.

34. X. Lin, R. Lu, and X. Shen, "Mdpa: multidimensional privacy-preserving aggregation scheme for wireless sensor networks," *Wireless Communications and Mobile Computing*, vol. 10, no. 6, pp. 843–856, 2010.

Chapter 3
An Efficient Authentication Scheme in Smart Grids

3.1 Introduction

Smart grids utilize information and communications technology to gather and act on information, such as information about the behavior of suppliers and consumers, in an automated fashion to improve the reliability, efficiency, economics, and sustainability of the generation and distribution of electricity [1, 2]. Smart grids are expected to be the next generation of power systems. In the traditional power grid, the architecture is featured with one-way electrical flows, i.e., electricity utility only delivers power to the consumers. In addition to the one-way electrical flows, smart grids also provide an attractive feature, i.e., two-way information flow communication. As depicted in Fig. 3.1, parallel to the one-way power flows, two-way information flows sharing is also implemented. In the two-way information flows sharing, the neighborhood gateway can collect electricity consumption reports from the customers via a wireless connection. Then, the neighborhood gateway sends the electricity reports to the control center via a wired link with high bandwidth and low delay. Based on the statistics and analysis of the above electricity reports, the control center can further convey the real-time pricing information to customers for their lower electricity bills, or send the control information to flatten demand peak [3,4].

Smart meter, the consumption-reporting device at the customer side, is vulnerable to malicious operations, e.g., the meter's reading modification [5]. Currently, in the US, such reading modifications have resulted in $6 billion loss [6]. It is indispensable for electricity utility to prevent malicious operations. Furthermore, with a mass of smart meters being deployed in smart grids, the malicious operations become more sophisticated. For instance, a large number of replayed/injected electricity consumption reports might maliciously be sent to the control center. If the attacks cannot be detected, the control center will be misled and make incorrect decisions, such as sending a false pricing information to the customers. Therefore, it is extremely important to develop an authentication scheme to detect the replayed/injected messages. In addition, a smart meter is only equipped with

H. Li, *Enabling Secure and Privacy Preserving Communications in Smart Grids*, 31
SpringerBriefs in Computer Science, DOI 10.1007/978-3-319-04945-8_3,
© The Author(s) 2014

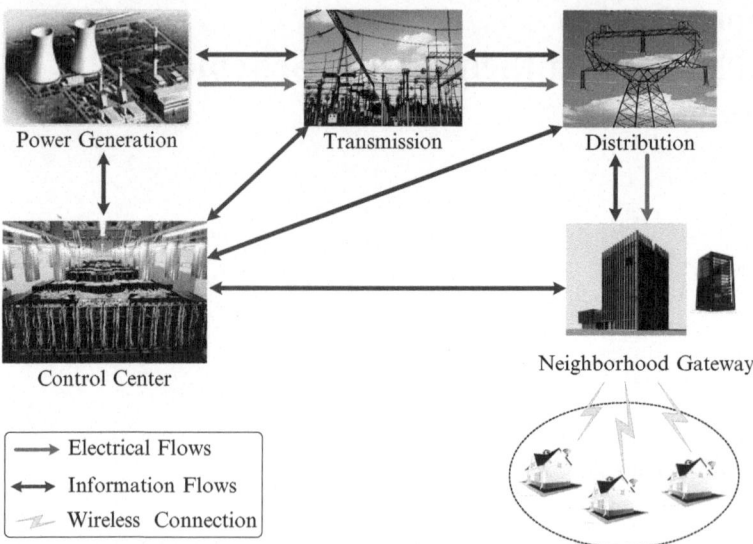

Fig. 3.1 Communication architecture for a smart grid

limited resources, i.e., a computation-constrained microprocessor, a small memory and a low computational capacity, etc. However, the computation overhead is heavy for the smart meter. For example, the initial deployments of the Advanced Metering Infrastructure (AMI) in Ontario, Canada, support meter readings at 5–60 min intervals [7]. The next generation of smart meters is planning to reduce these time intervals to 1 min or even less. Thus, the developed authentication scheme should minimize computation overhead on smart meters.

In this chapter, we propose an efficient authentication scheme to secure communication between the customers and the neighborhood gateway. Specifically, the contributions of this chapter are twofold. Firstly, we propose a novel authentication scheme, where the Merkle hash tree technique is leveraged to facilitate the authentication implementation. The security analysis indicates that the proposed scheme can resist the replay attack, the message injection attack, the message analysis attack, and the message modification attack. Secondly, extensive performance evaluation demonstrates that the proposed authentication scheme can achieve less communication overhead and dramatically reduce computation cost compared with the traditional authentication scheme, e.g., RSA-based authentication [8].

The remainder of the chapter is organized as follows. In Sect. 3.2, we present the network model, the threat model, and the design goal. Then, we propose the authentication scheme in Sect. 3.3, followed by the security analysis and performance evaluation in Sect. 3.4 and Sect. 3.5, respectively. We also present related works in Sect. 3.6. Finally, we draw our summary in Sect. 3.7.

3.2 Models and Design Goal

In this section, we present the network model, the threat model, and the design goal.

3.2.1 Network Model

As shown in Fig. 3.2, we consider a neighborhood area, which covers a neighborhood gateway connected with m Home Area Networks (HANs). The neighborhood gateway is equipped with a database for storing the detection and authentication information. The main functions of the neighborhood gateway are the replay attack detection, the message source authentication, and the message confidentiality and integrity guarantee. Each HAN is equipped with a smart meter, which contains a processor, non-volatile storage, and communication facilities. The smart meter can collect the electricity consumption report at small time-scales, e.g., every 15 min [7], and send the report to the neighborhood gateway. Note that the sent report is featured with a certain format. We assume that the neighborhood gateway knows the certain format, which can help the neighborhood gateway detect the report's integrity.

The neighborhood gateway is equipped with a high-power server, e.g., with the Inter Core i7 CPU and 6 GB Random Access Memory (RAM). The smart meter has limited resource, e.g., MSP430-F4270, 8 kB RAM and 120 kB flash memory [9]. For the smart meter, the computational efficiency should be considered, due to the limited computation resources. In addition, for the neighborhood gateway, since hundreds of electricity consumption reports will be synchronously collected, the computational efficiency is also a challenging issue. Meantime, communication between the HAN and the neighborhood gateway is through relatively inexpensive WiFi technology. Within the WiFi coverage of the neighborhood gateway, each HAN user can directly communicate with the neighborhood gateway.

3.2.2 Threat Model

We assume that both the neighborhood gateway and the HAN users cannot be compromised. We consider a global external adversary $\mathscr{A}$ as follows. *Global* indicates that the adversary $\mathscr{A}$ has full communication information of smart grids. *External* indicates that the adversary $\mathscr{A}$ can capture the communication messages between the neighborhood gateway and the HAN users, but not compromise the database of any HAN user. Specifically, we consider the adversary $\mathscr{A}$ can launch the following attacks.

- *Message analysis attack*: After eavesdropping a message from a HAN user, the external adversary $\mathscr{A}$ makes an attempt to recover the electricity consumer report. In this way, the privacy of the HAN user can be compromised.

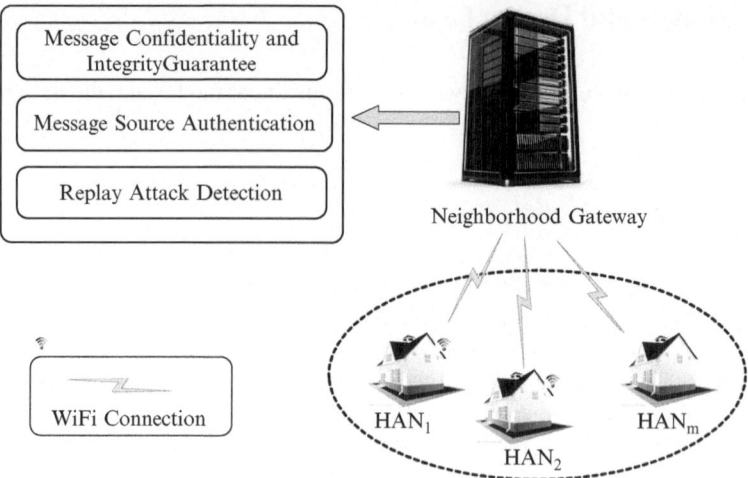

Fig. 3.2 Network model

- *Message modification attack*: The external adversary $\mathscr{A}$ captures a message from a HAN user, and attempts to tamper with the message.
- *Replay attack*: The external adversary $\mathscr{A}$ captures the previous message and replays the out-of-date messages to the neighborhood gateway.
- *Message injection attack*: The external adversary $\mathscr{A}$ sends fabricated messages to the neighborhood gateway. If the messages cannot be filtered, the control center might ultimately be mislead and make incorrect decisions. It is essential to develop an efficient authentication technique to identify the illegitimate message source.

3.2.3 Design Goal

Our design goal is to develop an efficient authentication scheme for smart grids. In specific, the following two desirable objectives should be achieved.

- The proposed scheme should achieve computation-effectiveness. The smart meter is equipped with a limited computation resource. And the computation overhead of the neighborhood gateway is also a challenging issue, since hundreds of HAN user electricity consumption reports will be synchronously collected to the neighborhood gateway. Therefore, the proposed scheme should consider the computation-effectiveness in both the smart meter and the neighborhood gateway.
- The proposed scheme should be secure in the threat model. As stated above, if the threats cannot be detected and excluded in smart grids, the performance of

smart grids will be degraded. Thus, the proposed scheme should resist the replay attack, the message injection attack, the message analysis attack, and the message modification attack.

3.3 Methodologies

In this section, we will propose our efficient authentication scheme, which consists of three phases: system initialization, reports generation, and neighborhood gateway authentication.

3.3.1 System Initialization

Based on the network model in Sect. 3.2.1, the following steps are performed to initialize the system.

- *Step-1*: For each HAN user $U_i (i = 1, \cdots, m)$, U_i securely communicates with the neighborhood gateway by running the Diffie-Hellman key establishment protocol [10]. Subsequently, a shared session key K_i is generated between U_i and the neighborhood gateway. Meantime, the symmetric encryption and decryption algorithms, e.g., AES [11], are shared by U_i and the neighborhood gateway. Let Enc and Dec denote the encryption and decryption algorithms, respectively.
- *Step-2*: As depicted in Fig. 3.3, each HAN user $U_i (i = 1, \cdots, m)$ constructs a Merkle hash tree with 128 leaf nodes in the following manner:

  - U_i randomly chooses 128 numbers $r_j \in \mathbb{Z}_q^* (j = 1, \cdots, 128)$, where r_j has the same size as the electricity consumption report. Then U_i picks 128 predefined timestamps $TS_j (j = 1, \cdots, 128)$. How to predefine timestamps will be described in *Step-3*.
  - U_i generates the AES ciphertext C_j with the session key K_i, where $C_j = Enc_{K_i}(r_j || TS_j)$ [11]. Then U_i computes the value of 128 leaf nodes, which are the cryptographic message hashes, $h_j = h(C_j)(j = 1, \cdots, 128)$, respectively.
  - U_i computes the value of internal nodes, which are derived from their child nodes. For example, the value of node $N_{1,2}$ is $h_{1,2} = h(h_1 | h_2)$ and $h_{127,128} = h(h_{127} | h_{128})$. Thus, U_i can recursively compute the values of the tree from the leaf nodes to the root node. Finally, the value of the root node is computed, $h_{1,128} = h(h_{1,64} | h_{65,128})$.

- *Step-3*: In order to achieve the real-time electricity consumption report collection, every t minutes, e.g., $t = 15$ min, each HAN user U_i should collect the electricity report and send it to the neighborhood gateway. Thus, there are 96 electricity report collections every day. Therefore, among the 128 leaf nodes, 96 leaf nodes are used as U_i's electricity report collection for every 15-min interval.

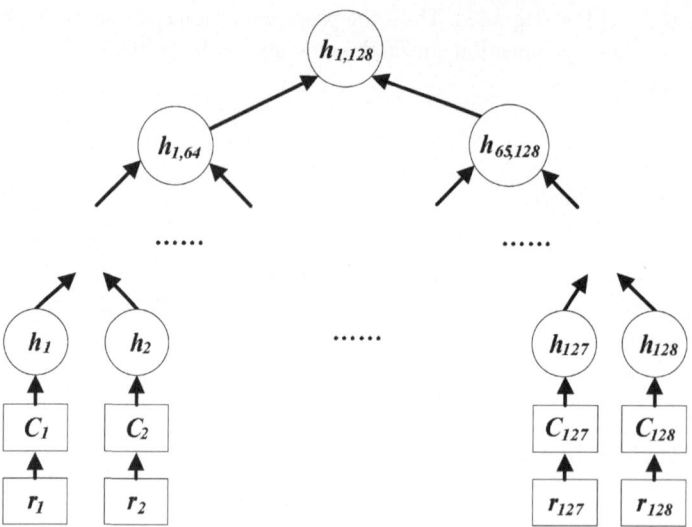

Fig. 3.3 System initialization

Table 3.1 HAN user electricity report collection

TS	r	C	API
0:00	r_1	C_1	$h_2, h_{3,4}, h_{5,8}, h_{9,16}, h_{17,32}, h_{33,64}, h_{65,128}$
0:15	r_2	C_2	$h_1, h_{3,4}, h_{5,8}, h_{9,16}, h_{17,32}, h_{33,64}, h_{65,128}$
0:30	r_3	C_3	$h_4, h_{1,2}, h_{5,8}, h_{9,16}, h_{17,32}, h_{33,64}, h_{65,128}$
...	...	...	...
23:45	r_{96}	C_{96}	$h_{95}, h_{93,94}, \cdots$
...	...	...	...
...	r_{128}	C_{128}	...

The other 32 leaf nodes are stored for other purposes, such as urgent electricity report collections. As we can see in Table 3.1, U_i creates a table in its local database. The table's data structure is in the form of (TS, r, C, API), where TS, r and C are the collection time of the electricity report, the corresponding random number and the corresponding ciphertext of r, respectively, API is the authentication path information. For example, at 0:00, r, C and API are r_1, C_1 and $< h_2, h_{3,4}, h_{5,8}, h_{9,16}, h_{17,32}, h_{33,64}, h_{65,128} >$, respectively.

- *Step-4*: U_i generates the ciphertext C of the root node value $h_{1,128}$ with the session key K_i [11], where $C = Enc_{K_i}(h_{1,128})$. Then U_i securely sends C to the neighborhood gateway.
- *Step-5*: After receiving C from U_i, the neighborhood gateway firstly decrypts C with the session key K_i to get the root value $h_{1,128}$, where $h_{1,128} = Dec_{K_i}(C)$. Then the neighborhood gateway creates a table in its local database. As shown in Table 3.2, the table's data structure is in the form of (ID, Root Value, Set of Hash(C_j)), where ID and Root Value are the identity of U_i and the hash tree's root value of U_i, respectively. The set of Hash(C_j) is the hash set of previously received C_j. The initialized set of Hash(C_j) is Null.

Table 3.2 HAN user authentication information

ID	Root value	Set of Hash(C_j)
U_1	$(h_{1,128})_{U_1}$	Null
U_2	$(h_{1,128})_{U_2}$	Null
U_3	$(h_{1,128})_{U_3}$	Null
...	...	...
U_m	$(h_{1,128})_{U_m}$	Null

Algorithm 1 Replay attack detection

1: **procedure** REPLAY ATTACK DETECTION
2:     According to ID = U_i, the neighborhood gateway searches Table 3.2 in the local database.
3:     **if** the returned result is Null **then**
4:         This message is damaged and the algorithm terminates. Then the neighborhood discards the message.
5:     **else**
6:         The corresponding set of Hash(C_j) is returned. Then the neighborhood gateway computes $h(C_j)$.
7:     **end if**
8:     **if** the returned set of Hash(C_j) is Null **then**
9:         The corresponding root value is returned and the algorithm terminates.
10:    **else if** $h(C_j)$ is the element of the returned set of Hash(C_j) **then**
11:        The replay attack is detected and the algorithm terminates.
12:    **else**
13:        The corresponding root value is returned and the algorithm terminates.
14:    **end if**
15: **end procedure**

3.3.2 Reports Generation

In order to achieve the real-time electricity consumption collection, every 15 min, HAN user $U_i (i = 1, \cdots , m)$ performs the following steps:

- *Step-1*: According to the current time TS_j, U_i searches Table 3.1 to get r_j, C_j and API_j. Then U_i uses the smart meters to collect the electricity consumption report D_j, and computes $S_j = r_j \oplus D_j$.
- *Step-2*: U_i sends the encrypted electricity consumption report $U_i||C_j||S_j||API_j$ to the neighborhood gateway.

3.3.3 Neighborhood Gateway Authentication

On receiving the encrypted electricity consumption report $U_i||C_j||S_j||API_j$, the neighborhood gateway performs the following steps:

- *Step-1*: *Replay attack detection.* With the detecting process in Algorithm 1, the neighborhood gateway can identify the replay attack.

Table 3.3 An example of detecting replay attack

ID	Root value	Set of Hash(C_j)
U_1	$(h_{1,128})_{U_1}$	Null
U_2	$(h_{1,128})_{U_2}$	d, e, f, g
U_3	$(h_{1,128})_{U_3}$	Null
...	...	...
U_m	$(h_{1,128})_{U_m}$	Null

An illustration example of Algorithm 1 is presented in Table 3.3. Let ID $= U_2$ and $h(C_j) = e$. Firstly, according to ID $= U_2$, the neighborhood gateway searches Table 3.3, and a matching ID can be found in Table 3.3. Then the corresponding set of Hash(C_j) is returned. Because the returned set of Hash(C_j) is $< d, e, f, g >$ which is not Null, the elements of $< d, e, f, g >$ will be compared with $h(C_j)$ one by one. Since e is already in the set of Hash(C_j), the report is a replayed one and will be discarded.

- *Step-2: Message source authentication.* On receiving the returned root value, the neighborhood gateway can authenticate the message source in conjunction with $h(C_j)$ and API_j. For example, the neighborhood gateway, which has stored the value of the root node $h_{1,128}$, needs to authenticate U_1 in conjunction with $h(C_1)$ and API_1, where $API_1 = < h_2, h_{3,4}, h_{5,8}, h_{9,16}, h_{17,32}, h_{33,64}, h_{65,128} >$. The neighborhood gateway can authenticate U_1 by in-turn computing

$$h_{1,2} = h(h(C_1)|h_2)$$
$$h_{1,4} = h(h_{1,2}|h_{3,4})$$
$$h_{1,8} = h(h_{1,4}|h_{5,8})$$
$$h_{1,16} = h(h_{1,8}|h_{9,16}) \tag{3.1}$$
$$h_{1,32} = h(h_{1,16}|h_{17,32})$$
$$h_{1,64} = h(h_{1,32}|h_{33,64})$$
$$h_{1,128} = h(h_{1,64}|h_{65,128})$$

Then the neighborhood gateway checks whether the computed $h_{1,128}$ is the same as the existing $h_{1,128}$. The neighborhood gateway accepts the report, only if the two values are equal. Finally, the neighborhood gateway inserts $h(C_j)$ into the set of Hash(C_j) of Table 3.2.

- *Step-3: Message confidentiality and integrity guarantee.* The neighborhood gateway performs the following operations:

- The neighborhood gateway runs AES decryption algorithm Dec to decrypt the received ciphertext C_j. With the session key K_i, the plaintext can be recovered by computing $Dec_{K_i}(C_j)$ [11]. The plaintext is in the form of $r_j||TS_j$.
- The neighborhood gateway checks the current local time TS' to detect whether the received message is fresh or not. We assume that θ is the predefined time

limit. Then the neighborhood gateway checks whether $TS' - TS_j \leq \theta$, if it holds, the neighborhood gateway accepts the fresh r_j, else discards r_j.

- The neighborhood gateway computes the U_i's electricity consumption report $D_j = r_j \oplus S_j$. It is easily verified that $D_j = r_j \oplus S_j = r_j \oplus (r_j \oplus D_j) = D_j$. As described in Sect. 3.2.1, the electricity report is featured with a certain format, which is also known by the neighborhood gateway. Thus, the neighborhood gateway can compare D_j's format with the existing one. If the comparison result is positive, the neighborhood gateway accepts the message, otherwise discards it.

3.4 Security Analysis

In this section, we analyze the security properties of the proposed authentication scheme. Especially, following the threat model discussed earlier, we are most concerned with how the proposed authentication scheme can resist the message analysis attack, the message modification attack, the replay attack, and the message injection attack.

3.4.1 Resist the Message Analysis Attack

Each HAH user U_i sends the encryption message C_j to the neighborhood gateway, where $C_j = Enc_{K_i}(r_j||TS_j)$ is the AES ciphertext encrypted by the session key K_j [11]. Since AES encryption algorithm is secure, the external adversary $\mathscr{A}$ cannot recover the plaintext $r_j||TS_j$ without the session key K_j. Thus, $\mathscr{A}$ cannot recover the electricity consumption report by computing $D_j = r_j \oplus S_j$. Therefore, the proposed authentication scheme can resist the message analysis attack.

3.4.2 Resist the Message Modification Attack

Each HAN user U_i sends a message to the neighborhood gateway in the form of $U_i||C_j||S_j||API_j$. After receiving the message, the neighborhood gateway firstly decrypts C_j with the session key K_i to recover the plaintext $r_j||TS_j$, then the neighborhood gateway computes the electricity consumption report $D_j = r_j \oplus S_j$. Furthermore, the neighborhood gateway checks the D_j's format. Because the format is certain, unless C_j and/or S_j were tampered by the external adversary $\mathscr{A}$, the format check will be positive. Therefore, by the format check, the proposed authentication scheme can resist the message modification attack.

3.4.3   Resist the Replay Attack

After receiving the message $U_i||C_j||S_j||API_j$, the neighborhood gateway runs Algorithm 1 to detect the replay attack. Firstly, the neighborhood gateway computes $h(C_j)$. Then according to the ID = U_i, the neighborhood gateway searches Table 3.2 and compares $h(C_j)$ with the elements of the returned set of Hash(C_j) one by one, if an equal value has existed in the set of Hash(C_j), the replay attack is detected. Because the set of Hash(C_j) covers all previous received $h(C_j)$, a message is considered as a replayed one if it is equal to any one of the set of Hash(C_j). Therefore, the proposed authentication scheme can resist the replay attack.

3.4.4   Resist the Message Injection Attack

After the replay attack detection, the neighborhood gateway will authenticate the message source. As described in Sect. 1.3.4, With the $h(C_j)$, API_j and $h_{1,128}$, the neighborhood gateway can in-turn compute hash values along the path from the leaf node to the root node. In the end, $h_{1,128}$ can be computed. Then the computed $h_{1,128}$ is compared to the existing $h_{1,128}$. If the two values are equal, the message source is considered to be a legitimate HAN user, otherwise a message injection attack is detected. Since the external adversary $\mathscr{A}$ cannot compromise the database of the HAN user to steal the secret information and the Merkle hash tree authentication is proved to be secure [12], $\mathscr{A}$ cannot pass the authentication process.

Note that if an external adversary $\mathscr{A}$ compromises the communication between U_i and the neighborhood gateway and captures the message $U_i||C_j||S_j||API_j$, $\mathscr{A}$ can fabricate a message, denoted by C_j' so that $h(C_j') = h(C_j)$—which is called a collision—then sends the fabricated message in the form of $U_i||C_j'||S_j||API_j$. As stated above, this type of attack cannot be detected. In the following, we will prove the probability of $h(C_j') = h(C_j)$ is negligible.

In the proposed authentication scheme, the hash function $h(.)$ adopts a 128-bit cryptographic hash function [13]. Therefore, for a random message m, there are 2^{128} possible values for $h(m)$. Here, we demonstrate how many fabricated messages have been delivered to the neighborhood gateway when the collision, i.e., $h(C_j') = h(C_j)$ occurs. We assume that $P(n)$ is the probability that more than one collision occurs after n fabricated messages were sent to the neighborhood gateway. Let E_j denote the event that the jth fabricated message collides with the $h(C_j)$, Then, $Pr[E_j] = (j-1)/2^{128}$, and

$$
\begin{aligned}
P(n) &= Pr[E_1 \vee E_2 \vee \cdots \vee E_n] \\
&\leq Pr[E_1] \vee Pr[E_2] \vee \cdots \vee Pr[E_n] \\
&\leq \frac{0}{2^{128}} + \frac{1}{2^{128}} + \cdots + \frac{n-1}{2^{128}} \\
&= \frac{n(n-1)}{2^{129}}
\end{aligned}
\tag{3.2}
$$

As shown in Eq. (3.2), the upper bound of the collision probability $P(n)$ grows with $O(2^{-129}n^2)$. When $P(n) \to 1/2, n^2 \approx 2^{128}$. Thus, $n \approx 2^{64}$. Therefore, after at least 2^{64} fabricated messages were launched to the neighborhood gateway, a collision might occur with 50% chance. Note that the C_j varies every 15 min. As a result, in the limited time period, the collision probability is negligible.

In summary, the proposed authentication scheme can resist the message injection attack.

3.5 Performance Evaluation

In this section, we focus on investigating the authentication performance. Since the RSA algorithm was suggested to secure smart grids [8], we compare the proposed authentication scheme with the RSA-based authentication scheme in terms of the communication overhead between the HAN user and the neighborhood gateway, and the computation complexity of the HAN user and the neighborhood gateway.

3.5.1 Communication Overhead

In the RSA-based authentication scheme, the HAN user sends an RSA signature to neighborhood gateway. Considering the popular security, we choose 1,024 bits as the size of an RSA signature. In comparison, in the proposed authentication scheme, the HAN user sends the message to the neighborhood gateway in the form of $U_i||C_j||S_j||API_j$. Note that $U_i||C_j||S_j$ is served for detecting the replay attack and assuring message confidentiality and integrity, only API_j is served for authenticating the message source. As stated in Table 3.1, API_j includes 7 128-bit cryptographic hash values, thus the total communication overhead is $128 \times 7 = 896$ bits.

Figure 3.4 depicts the communication overhead at a given neighborhood gateway for different number of HAN users. When the number of HAN users is small, the communication overhead is low in both the proposed scheme and the RSA-based scheme. Then the communication overhead increases with the increased HAN users. However, it should be noted that the increase is faster in the case of the RSA-based scheme. For example, when 2,000 HAN users are considered for the given neighborhood gateway, the communication overhead of the RSA-based scheme is 2,000 kb, but only 1,750 kb in the proposed scheme. The RSA-based scheme bears higher communication overhead because of the RSA signature included in the message.

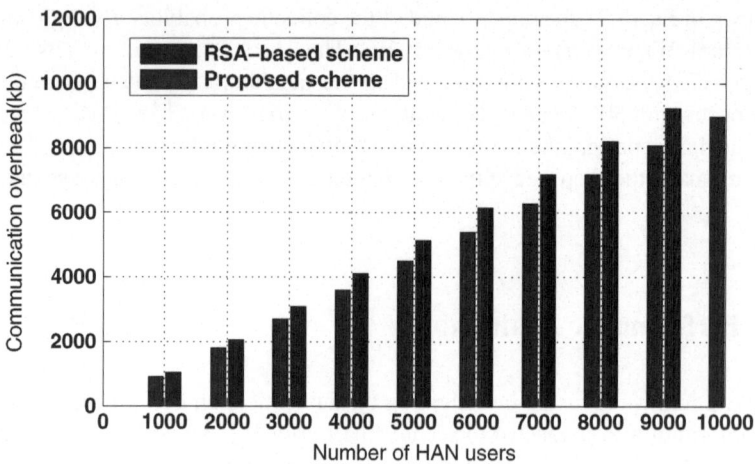

Fig. 3.4 Comparison of communication overhead

Table 3.4 Execution time of cryptographic operations

	Descriptions	Execution time
T_h	The time of one cryptographic hash	0.000095 ms
T_s	The time of one RSA signature	2.26 ms
T_v	The time of one RSA signature verification	0.11 ms

3.5.2 Computation Complexity

For both the proposed authentication scheme and the RSA-based scheme, since the RSA signing, RSA signature verifying, and the cryptographic hash computations dominate the computation complexity, we only count the number of these operations in the assessment of computation performance.

In the RSA-based authentication scheme, the HAN user needs to perform an RSA signature to sign the electricity consumption. For the neighborhood gateway, it performs an RSA signature verification to verify the signature. However, in the proposed authentication scheme, since the HAN user only performs electricity consumption report collection and exclusive-OR operation, the computation complexity is negligible. On the other hand, for the neighborhood gateway, it costs seven hashes to compute the hash value of the root node.

We conduct experiments to study the execution time. Table 3.4 gives the observed processing time. The implementation was executed on an Intel Pentium IV 3.0 GHz machine [14]. Figure 3.5 shows the computation costs of the HAN users. As we can see, in the RSA-based authentication scheme, with the increasing number of HAN users, the total computation cost significantly increases. On the contrary, in the proposed scheme, the computation cost is constantly low. For instance, when the number of HAN users hits 4,000, the computation cost of the RSA-based scheme is relatively high (9,000 ms) in contrast with a significantly low value (near to zero) for the proposed scheme. The RSA-based scheme experiences higher computation cost due to the time-consuming RSA signature computation. Therefore, the proposed scheme enables the resource-restrained HAN user to be more lightweight. Figure 3.6 shows the computation cost of the neighborhood gateway for different number of HAN users. It can be easily seen that the proposed scheme achieves a lower computation cost. For example, when the number of HAN users reaches 4,000, the computation cost of the proposed scheme is only 0.36 ms, but 400 ms in the RSA-based scheme. The proposed scheme experiences lower computation cost due to the efficient Merkle hash tree authentication. Thus, the proposed scheme achieves more computation-efficient neighborhood gateway than the RSA-based scheme.

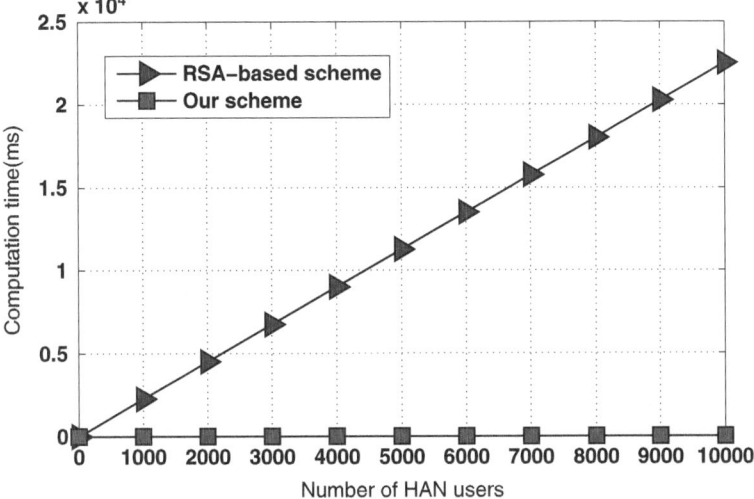

Fig. 3.5   Computation cost of HAN users

3.6   Related Works

Merkle hash tree [12] has been applied in many research works [15, 16]. These works have led to the following authentication schemes [17, 18]. Xu et al. [17] proposed a hash tree-based authentication scheme in Session Initiation Protocol (SIP). The scheme can be used in SIP entities which have less computation power and limited memory. Lin and Sung [18] developed an efficient source authentication scheme for multicast based on the Merkle hash tree. The scheme can reduce both communication cost and computation cost compared to other source authentication schemes. On the other hand, authentication is also essential for smart grids. Because not all the entities in smart grids are trusted, for a secure smart grid communication,

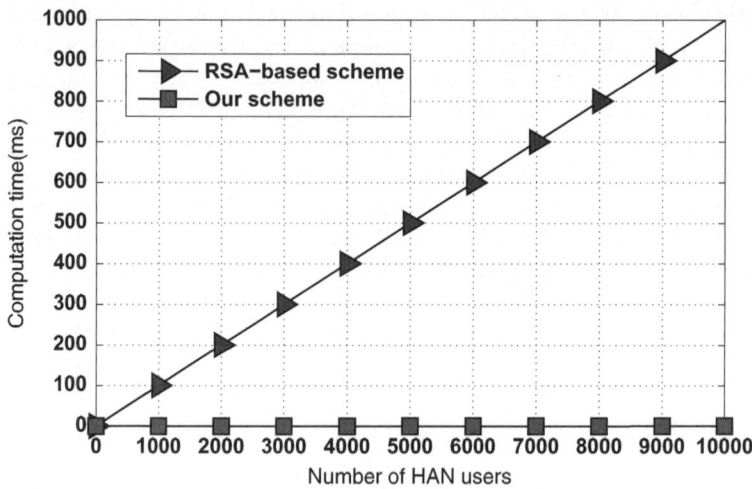

Fig. 3.6 Computation cost of the neighborhood gateway

it is indispensable to verify whether the parties involved in communication are the exact entities they appear to be. Therefore, the authentication technique should be developed so that the attackers cannot impersonate legitimate entities in smart grids. Hamlyn et al. [19] proposed a new utility computer network security management and authentication, which can be used for action or command requests in smart grid operations. Ayday and Rajagopal [20] proposed three secure, intuitive, and low-cost device authentication mechanisms for the HAN part of smart grid networks. These mechanisms are resilient to adversarial behavior including man-in-the-middle and impersonation attacks. Li and Cao [21] developed a multicast authentication in smart grids with a one-time signature. The scheme can reduce storage cost and signature size, and flexibly allocate the computations between the sender and receiver based on their computing resources. Fouda et al. [22] proposed a message authentication scheme for smart grid communications. Based on the Diffie-Hellman key establishment protocol, the scheme allows smart meters to make mutual authentication and achieve message authentication with low latency and a few exchanges of signal messages.

3.7 Summary

In this chapter, we have proposed an efficient authentication scheme tailored to the requirements of secure smart grids. Detailed security analysis shows that the proposed authentication scheme can resist the replay attack, the message injection attack, the message analysis attack, and the message modification attack. Extensive performance evaluation further demonstrates its efficiency in terms of computation complexity and communication overhead.

References

1. H. Liang, B. Choi, W. Zhuang, and X. Shen, "Towards optimal energy store-carry-and-deliver for phevs via v2g system," in *IEEE INFOCOM*, 2012, pp. 1674–1682.
2. H. Liang, B. Choi, A. Abdrabou, W. Zhuang, and X. Shen, "Decentralized economic dispatch in microgrids via heterogeneous wireless networks," *IEEE Journal on Selected Areas in Communications*, vol. 30, no. 6, pp. 1061–1074, 2012.
3. H. Liu, H. Ning, Y. Zhang, and L. Yang, "Aggregated-proofs based privacy-preserving authentication for v2g networks in smart grid," *IEEE Transactions on Smart Grid*, vol. 3, no. 4, pp. 1722–1733, 2012.
4. H. Li, X. Liang, R. Lu, X. Lin, and X. Shen, "Edr: An efficient demand response scheme for achieving forward secrecy in smart grid," in *IEEE GLOBECOM*, 2012, pp. 929–934.
5. M. Wen, R. Lu, K. Zhang, J. Lei, X. Liang, and X. Shen, "Parq: A privacy- preserving range query scheme over encrypted metering data for smart grid," *IEEE Transactions on Emerging Topics in Computing*, vol. 1, no. 1, pp. 178–191, 2013.
6. P. McDaniel and S. McLaughlin, "Security and privacy challenges in the smart grid," *IEEE Security & Privacy*, vol. 7, no. 3, pp. 75–77, 2009.
7. A. Cavoukian, "Privacy by design," *Information and Privacy Commissioner of Ontario, Canada*, 2009.
8. "Introduction to nistir 7628 guidelines for smart grid cyber security," *Available: http://www.nist.gov/smartgrid/upload//nistir-7628_total.pdf*, 2010.
9. D. O'Connell and I. De Vries, "Digital energy metering for electrical system management," in *ACM Symposium on Applied Computing*, 2010, pp. 516–520.
10. D. Stinson, *Cryptography: theory and practice*. CRC press, 2006.
11. N. Ferguson, R. Schroeppel, and D. Whiting, "A simple algebraic representation of rijndael," in *Selected Areas in Cryptography*, 2001, pp. 103–111.
12. R. Merkle, "Protocols for public key cryptosystems," in *IEEE Symposium on Security and privacy*, 1980, pp. 122–134.
13. R. Lu, X. Lin, H. Zhu, P. Ho, and X. Shen, "A novel anonymous mutual authentication protocol with provable link-layer location privacy," *IEEE Transactions on Vehicular Technology*, vol. 58, no. 3, pp. 1454–1466, 2009.
14. W. Dai, "Crypto++ 5.6.0 benchmarks," http://www.cryptopp.com/benchmarks.html, 2009.
15. B. Przydatek, D. Song, and A. Perrig, "Sia: Secure information aggregation in sensor networks," in *ACM conference on Embedded networked sensor systems*, 2003, pp. 255–265.
16. K. Ren, W. Lou, K. Zeng, and P. Moran, "On broadcast authentication in wireless sensor networks," *IEEE Transactions on Wireless Communications*, vol. 6, no. 11, pp. 4136–4144, 2007.
17. K. Xu, X. Ma, and C. Liu, "A hash tree based authentication scheme in sip applications," in *International Conference on Communications*, 2008, pp. 1510–1514.
18. I. Lin and C. Sung, "An efficient source authentication for multicast based on merkle hash tree," in *the Sixth International Conference on Intelligent Information Hiding and Multimedia Signal Processing*, 2010, pp. 5–8.
19. A. Hamlyn, H. Cheung, T. Mander, L. Wang, C. Yang, and R. Cheung, "Network security management and authentication of actions for smart grids operations," in *IEEE Canada Electrical Power Conference*, 2007, pp. 31–36.

20. E. Ayday and S. Rajagopal, "Secure, intuitive and low-cost device authentication for smart grid networks," in *IEEE Consumer Communications and Networking Conference (CCNC)*, 2011, pp. 1161–1165.
21. Q. Li and G. Cao, "Multicast authentication in the smart grid with one-time signature," *IEEE Transactions on Smart Grid*, vol. 2, no. 4, pp. 686–696, 2011.
22. M. Fouda, Z. Fadlullah, N. Kato, R. Lu, and X. Shen, "A lightweight message authentication scheme for smart grid communications," *IEEE Transactions on Smart Grid*, vol. 2, no. 4, pp. 675–685, 2011.

Chapter 4
An Efficient Fine-Grained Keywords Comparison Scheme in the Smart Grid Auction Market

4.1 Introduction

With the increasing requirements of electricity consumption and the wide application of renewable energy sources, the traditional power grid cannot sustain the development trend. It is restructured to more intelligent power systems called smart grids [1, 2]. Smart grids mainly consist of several parts: generator(s), transmission system operator, distributor(s), retailer(s) and aggregator(s) [3]. Many technologies have been introduced into smart grids to ensure their availability and economic benefits [4]. For instance, the energy auction market introduces commercial auctions to smart grids, where energy sellers publish their auction information, and then energy buyers bid for appropriate energy supplies. Thus, the energy auction market can adjust energy prices and provide strong support for the practical application of smart grids [5].

However, in the energy auction market, security and privacy are seriously challenged. Firstly, due to the particularity of the energy auction market, privacy preservation is extremely important because auction information is closely related to trade secrets. In addition, many traditional security requirements are still needed: sensitive information should be encrypted, legitimate users should be authenticated, and illegal users cannot modify communication messages.

In this chapter, we propose an efficient fine-grained keywords comparison (EFKC) scheme in the smart grid auction market. This scheme focuses on providing secure and efficient transactions between generators and retailers. It considers auction message encryption, keyword search, fine-grained comparison and auction message pre-filtering. Specifically, the contributions of this chapter are twofold.

- Firstly, we propose a novel EFKC scheme to achieve searchable encryption. This scheme not only compares whether the keywords are equal to each other, but also accurately calculates the difference between multidimensional keywords, which is more practical in an auction market. Also, security analysis demonstrates

H. Li, *Enabling Secure and Privacy Preserving Communications in Smart Grids*, 47
SpringerBriefs in Computer Science, DOI 10.1007/978-3-319-04945-8_4,
© The Author(s) 2014

that the EFKC scheme can achieve privacy preservation, authentication, data integrity, and confidentiality.

* Secondly, based on the two super-increasing sequences, the proposed scheme can compare the multidimensional keywords of one buyer with those of all sellers with only one calculation, thereby greatly reducing the computation and communication overhead. We compare the EFKC scheme with the auction scheme [4] to show its efficiency.

The remainder of the chapter is organized as follows.In Sect. 4.2, the system model, security requirements, and design goals are formalized. In Sect. 4.3, we propose the EFKC scheme. We analyze security of our scheme in Sect. 4.4, and evaluate its performance in Sect. 4.5. In Sect. 4.6, we present related works. Finally, we conclude this chapter in Sect. 4.7.

4.2 System Model, Security Requirements, and Design Goals

In this section, we will formalize the system model, security requirements, and design goals.

4.2.1 System Model

In our system model, we focus on how to secretly compare keyword tags with trapdoors generated by sellers and buyers in the energy market. Specifically, we consider that our system consists of four parts, as shown in Fig. 4.1.

* *Electricity generators (sellers)*: Sellers generate energy and sell it to retailers. Hence they generate their auction messages and make the corresponding keyword tags for efficient search, then send them to the data center.
* *Retailers (buyers)*: Buyers should provide energy to their own energy consumers. For economic purposes, they generate keyword trapdoors to filter the auction messages and send them to the data center.
* *Data Center (DC)*: The data center in our scheme acts as a database, it stores all tags and auction messages from sellers. If one buyer computes a trapdoor to bid some auction messages, DC will compare the trapdoor with all tags through homomorphic computing. Then, DC sends the result to the filtering center.
* *Filtering Center (FC)*: The filtering Center is a trusted operation center that may be a supercomputer. It initiates the whole system at the beginning of the auction. And after the comparison, the FC needs to recover the corresponding result and communicate with the DC to pre-filter the auction messages. Finally, FC chooses the winners and sends them to the buyers.

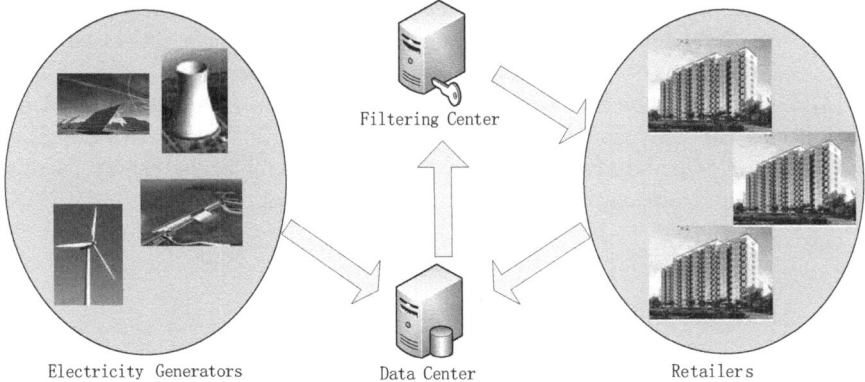

Fig. 4.1 System model for the smart grid auction market

4.2.2 Security Requirements

In our scheme, we assume all entities are untrustworthy except FC. Thus the adversary $\mathscr{A}$ can intrude in smart grids and eavesdrop on the messages with private information. Therefore, we define our security requirements against the malicious behavior.

- *Privacy preservation of auction messages and their keywords*: The auction messages and all keywords generated by sellers and buyers must be sent to DC for comparing and filtering. However, these information may be trade secrets. Hence, it is most important to guarantee privacy preservation even though the adversary $\mathscr{A}$ eavesdrops on the messages or DC's database.
- *Authentication and data integrity*: In the system, legitimate users should be authenticated, and the messages altered or fabricated by adversary $\mathscr{A}$ should be detected.
- *Confidentiality*: When the message contains sensitive information, it is indispensable to encrypt it. We should ensure that only the receiver can decrypt it, and the adversary $\mathscr{A}$, even though other untrustworthy entities cannot identify it.

4.2.3 Design Goals

In order to realize the auction messages filtering in our scheme, our design goals are to develop an EFKC with privacy preservation.

- *Security is indispensable in the proposed scheme*: If the auction market in smart grids doesn't consider the security, it cannot be used in practice. Hence, we should guarantee privacy preservation of auction messages and keywords, authentication, data integrity, and confidentiality.

- *Computation and communication efficiency should be achieved in the proposed scheme*: Compared with other auction schemes, our scheme should be more effective in terms of computation and communication overhead.
- *Keywords should be fine-grained compared in the proposed scheme*: General schemes can only compare whether the keywords are equal. However, the difference of keywords computing will be very useful in the energy auction market. Therefore, our scheme should achieve this goal.

4.3 Methodologies

In this section, we propose our EFKC scheme, which mainly consists of the following three phases: system initialization, auction message creating, and filtering.

4.3.1 System Initialization

4.3.1.1 Identity-Based Signature and Signcryption Algorithm

At the beginning of system establishment, given a security parameter κ, FC firstly generates $(q, P, \mathbb{G}_1, \mathbb{G}_2, e)$ by running $\mathscr{G}en(\kappa)$, then chooses $s \in \mathbb{Z}_q^*$ and computes the global public key $P_{pub} = sP$. Every entity should register itself to the system, and get a pair of identity-based public/private key from the FC, e.g., the FC's public key is $P_{FC} = H_0(ID_{FC})$, where $ID_{FC} \in \{0, 1\}^{k_0}$ is the identity of the FC (k_0 is the number of bits required to represent an identity), and the corresponding private key is $S_{FC} = sP_{FC}$. Similarly, the *seller$_i$*, *buyer$_j$* and data center(DC) get their public/private key (P_{S_i}, S_{S_i}), (P_{B_i}, S_{B_i}) and (P_{DC}, S_{DC}) from the FC, respectively. We also need two hash functions $H_0, H_1 : \{0, 1\}^* \rightarrow \mathbb{G}_1^*$. These parameters will support the identity-based signature algorithm [6]. Finally, we use signcryption algorithm [7] to signcrypt some messages in Sect. 4.3.3. Due to the limitation of length, no more tautology here.

4.3.1.2 Two Super-Increasing Sequences for Aggregation

Firstly, FC computes the Paillier Cryptosystem's public key (n, g) and the corresponding private key (λ, μ). Taking into account the multidimensional keywords of auction information, we expect that all keywords can be aggregated to one number and the difference of all keywords can be gained by only one comparison. All keywords can be transformed to a corresponding positive integer. Therefore, we transform each dimension keyword to a positive integer. Assume that for *seller$_i$* there are totally l types of auction keywords $(m_{i,1}, m_{i,2}, \cdots, m_{i,l})(m_{i,j} \in \mathbb{Z}_n)$, and the value of each type $m_{i,j}$ $(j = 1, 2, \cdots, l)$ is less than a constant d. Then, the FC chooses a super-increasing sequence $\mathbf{a} = (a_1 = 1, a_2, \cdots, a_l)$, where $a_1, a_2, \cdots, a_l$ are large integers and $\sum_{j=1}^{i-1} a_j \cdot d < a_i/2$ for $i = 2, \cdots, l$. The reason why we choose $a_i/2$ will be described in Sect. 4.3.3. Then, the FC computes $(g_1, g_2, \cdots, g_l)$, where $g_i = g^{a_i}$ $(i = 1, 2, \cdots, l)$.

Then we define the *seller*'s aggregated number of multidimensional keywords is no more than a constant D, e.g., $\sum_{j=1}^{l} a_j \cdot d < D$, and the FC further chooses another super-increasing sequence $\mathbf{b} = (b_1 = 1, b_2, \cdots, b_I)$ (I is the number of *sellers*), where $\sum_{j=1}^{i-1} b_j \cdot D < b_i/2$. The reason why we choose $b_i/2$ will be also described in Sect. 4.3.3.

After all, the FC publishes the system parameters as

$$\text{pubs} = \left\{ \begin{array}{l} q, P, P_{pub}, \mathbb{G}_1, \mathbb{G}_2, e, k_0, H_0, \\ H_1, n, g, g_1, g_2, \cdots, g_l, \mathbf{b} \end{array} \right\} \tag{4.1}$$

and keeps the master keys $(\lambda, \mu, \mathbf{a}, s)$ secretly.

4.3.2 Auction Message Creating

4.3.2.1 Auction Creating

In this phase, *seller$_i$* firstly generates his auction message $auc_i \in \mathbb{Z}_n$ and computes $A_i = g^{auc_i} \cdot r_i^n \bmod n^2$, where $r_i \in \mathbb{Z}_n^*$ is a random number.

4.3.2.2 Tag Creating

seller$_i$ selects keywords $(m_{i,1}, m_{i,2}, \cdots, m_{i,l})$ according to the corresponding auc_i, then computes his tag

$$\begin{aligned} C_i &= (g_1^{m_{i,1}} \cdot g_2^{m_{i,2}} \cdot \cdots \cdot g_l^{m_{i,l}} \cdot r_i^n)^{b_i} \bmod n^2 \\ &= g^{(a_1 m_{i,1} + a_2 m_{i,2} + \cdots + a_l m_{i,l})b_i} \cdot (r_i^{b_i})^n \bmod n^2 \\ &= g^{b_i M_i} \cdot (r_i^{b_i})^n \bmod n^2 \end{aligned} \tag{4.2}$$

where $M_i = a_1 m_{i,1} + a_2 m_{i,2} + \cdots + a_l m_{i,l}$.

4.3.2.3 Delivery

seller$_i$ signs the message $msg_i = (A_i || C_i || ID_{S_i} || TS)$ with his own private key S_{S_i} as follows: *seller$_i$* randomly chooses $x \in \mathbb{Z}_q$, and computes

$$
\begin{cases}
U = xP \\
V = S_{S_i} + xH_1(ID_{S_i}, msg_i, U)
\end{cases}
\tag{4.3}
$$

Then the corresponding signature is $\sigma = <U, V>$. $seller_i$ sends the signed message $(msg_i \| \sigma)$ to DC. After receiving the message $(msg_i \| \sigma)$ from $seller_i$, DC accepts it if

$$
e(P, V) = e(P_{pub}, P_{S_i}) e(U, H_1(ID_{S_i}, msg_i, U))
\tag{4.4}
$$

4.3.3 Filtering

4.3.3.1 All Sellers' Tags Aggregation

After receiving all *sellers'* tags, DC computes the total tag as $C_{total} = C_1 \cdot C_2 \cdots C_I$, where

$$
\begin{aligned}
C_{total} &= C_1 \cdot C_2 \cdots C_I \\
&= \prod_{i=1}^{I}(g^{b_i M_i}) \cdot (\prod_{i=1}^{I} r_i^{b_i})^n \bmod n^2 \\
&= g^{\sum_{i=1}^{I} b_i M_i} \cdot (\prod_{i=1}^{I} r_i^{b_i})^n \bmod n^2
\end{aligned}
\tag{4.5}
$$

4.3.3.2 Trapdoor Creation and Delivery

When $buyer_j$ wants to bid the energy, he firstly selects keywords $(m_{j,1}, m_{j,2}, \cdots, m_{j,l})$ and randomly chooses $r_j \in \mathbb{Z}_n^*$, then computes his trapdoor

$$
\begin{aligned}
C_j' &= g_1^{-m_{j,1}} \cdot g_2^{-m_{j,2}} \cdots g_l^{-m_{j,l}} \cdot r_j^n \bmod n^2 \\
&= g^{-M_j} \cdot r_j^n \bmod n^2
\end{aligned}
\tag{4.6}
$$

where $M_j = a_1 m_{j,1} + a_2 m_{j,2} + \cdots + a_l m_{j,l}$. And then $buyer_j$ calculates the total trapdoor as $C_{total}' = C_j'^{b_1 + b_2 + \cdots + b_I}$, where

$$
\begin{aligned}
C_{total}' &= C_j'^{b_1 + b_2 + \cdots + b_I} \\
&= (g^{-M_j} \cdot r_j^n)^{b_1 + \cdots + b_I} \bmod n^2 \\
&= g^{-\sum_{i=1}^{I} b_i M_j} \cdot (\prod_{i=1}^{I} r_j^{b_i})^n \bmod n^2
\end{aligned}
\tag{4.7}
$$

After that, $buyer_j$ signs $msg_j = (C'_{total}||ID_{B_j}||TS)$ as Eq. (4.3) with his own private key. Then he sends the signed message to DC, and the DC verifies it as Eq. (4.4).

4.3.3.3 Homomorphic Computing for Comparison

When the DC wants to compare *sellers'* tags with $buyer_j$'s trapdoor, it can compute $C = C_{total} \cdot C'_{total}$. Then the DC signs the message $(C||ID_{DC}||TS)$ as Eq. (4.3) with its own private key.

4.3.3.4 Decrypting the Result of Comparison

After the signature verification, the FC decrypts C, where C is formed by

$$
\begin{aligned}
C &= C_{total} \cdot C'_{total} \\
&= g^{\sum_{i=1}^{I} b_i M_i} \cdot (\prod_{i=1}^{I} r_i^{b_i})^n \cdot g^{-\sum_{i=1}^{I} b_i M_j} \cdot (\prod_{i=1}^{I} r_j^{b_i})^n \bmod n^2 \\
&= g^{\sum_{i=1}^{I} b_i M_{i,j}} \cdot (\prod_{i=1}^{I} (r_i r_j)^{b_i})^n \bmod n^2 \\
&= g^{M_{total}} \cdot (\prod_{i=1}^{I} (r_i r_j)^{b_i})^n \bmod n^2
\end{aligned}
\tag{4.8}
$$

where $M_{i,j} = M_i - M_j = a_1(m_{i,1} - m_{j,1}) + a_2(m_{i,2} - m_{j,2}) + \cdots + a_l(m_{i,l} - m_{j,l})$ and $M_{total} = \sum_{i=1}^{I} b_i M_{i,j}$. The FC uses the $sk(\lambda, \mu)$ to recover M_{total}. After that FC gets $(M_{1,j}, M_{2,j}, \cdots, M_{I,j})$ by running Algorithm 1 with input $\mathbf{x} = \mathbf{b}$ and $SUM = M_{total}$.

As shown in Algorithm 1, we define $sum_i = b_1 M_{1,j} + b_2 M_{2,j} + \cdots + b_i M_{i,j} (i = 1, 2, \cdots, I)$. We compute $sum_{i-1} = sum_i \bmod x_i$, hence we have $0 \leq sum_{i-1} < x_i$. Since we have defined $\sum_{j=1}^{i-1} b_j \cdot D < b_i/2$, we have $-x_i/2 < sum_{i-1} < x_i/2$ (for example: $0 < \varepsilon_i, \varepsilon_j < t \implies -t < \varepsilon_i - \varepsilon_j < t$). Thus in Algorithm 1 if the calculated sum_{i-1} is $0 \leq sum_{i-1} < x_i/2$, this is the right result; else if $x_i/2 < sum_{i-1} < x_i$, we must correct it as $sum_{i-1} = sum_{i-1} - x_i$, the true result is $-x_i/2 < sum_{i-1} < 0$. That's why we choose $b_i/2$ in $\sum_{j=1}^{i-1} b_j \cdot D < b_i/2$ and $a_i/2$ in $\sum_{j=1}^{i-1} a_j \cdot d < a_i/2$, it can split the aggregation including negative numbers.

After getting $(M_{1,j}, M_{2,j}, \cdots, M_{I,j})$, the FC can use Algorithm 1 with input $\mathbf{x} = \mathbf{a}$ and $SUM = M_{i,j} (i = 1, 2, \cdots, I)$ to gain all difference of the multidimensional keywords $(dif_1, dif_2, \cdots, dif_l)$.

Algorithm 1 Split the aggregation

Input: $\mathbf{x} = (x_1, x_2, \cdots, x_k)$ and the aggregation SUM
Output: $(D_1, D_2, \cdots, D_k)$
1: Let $sum_k = SUM$
2: **for** $i = k$ to 2 **do**
3:      $sum_{i-1} = sum_i \bmod x_i$
4:      **if** $sum_{i-1} > x_i/2$ **then**
5:          $sum_{i-1} = sum_{i-1} - x_i$
6:      **end if**
7:      $D_i = \frac{sum_i - sum_{i-1}}{x_i}$
8: **end for**
9: $D_1 = sum_1$
10: **return** $(D_1, D_2, \cdots, D_k)$

4.3.3.5   Getting Pre-filtering Auctions

If the result of comparison between $seller_i$'s tag and $buyer_j$'s trapdoor satisfies the pre-filtering rules (e.g., if we need the i-dimensional keywords of tag and trapdoor are same, the rule should be $dif_i = 0$; if we need the tag's j-dimensional keyword is more than trapdoor's, the rule should be $dif_j > 0$; ...), the FC stores the corresponding ID_{S_i} in an array $sellers[]$. To send the message $(sellers[]||ID_{FC}||TS)$ to the DC, the FC signcrypts the message with the signcryption algorithm [7], and sends the ciphertext to the DC. After receiving the ciphertext, the DC recovers the $sellers[]$.

Then the DC selects the corresponding A_i according to $sellers[]$ and stores them in a pre-filtering array $A[]$. The DC signs the message $(A[]||ID_{DC}||TS)$ and sends it to the FC, the FC accepts it after verification. Finally, the FC chooses winners by some selection criterion [4]. Since winner selection criterion is not the focus of our research, we do not consider it in detail. The selected winners will be stored in an array list $win[]$ and sent to $buyer_j$ after signcrypting [7].

4.4   Security Analysis

In this section, we will analyze the security properties of our proposed scheme. In particular, based on the security requirements discussed above, our analysis will focus on how the proposed scheme can achieve fine-grained keywords comparison secretly, privacy preservation, authentication, data integrity, and confidentiality.

4.4.1 Secure Fine-Grained Keywords Comparison Between Tag and Trapdoor

In our proposed scheme, the tag's all types of keywords $(m_{i,1}, m_{i,2}, \cdots, m_{i,l},)(m_{i,j} \in \mathbb{Z}_n)$ are aggregated to C_i as

$$
\begin{aligned}
C_i &= (g_1^{m_{i,1}} \cdot g_2^{m_{i,2}} \cdots \cdots g_l^{m_{i,l}} \cdot r_i^n)^{b_i} \bmod n^2 \\
&= g^{b_i M_i} \cdot (r_i^{b_i})^n \bmod n^2
\end{aligned}
\tag{4.9}
$$

This means that C_i is a ciphertext of the Paillier cryptosystem. Similarly, C_j, C_{total}, and C'_{total} are also the same. Due to the semantic security of the Paillier cryptosystem, the privacy preservation of keywords is achieved. If the adversary $\mathscr{A}$ eavesdrops on them, he cannot identify the tag or the trapdoor. And in DC, since it only does homomorphic computing on C_{total} and C'_{total}, it cannot identify the tag or trapdoor. In the end, the FC will decrypt C to get the fine-grained comparison results of keywords; the FC still cannot gain each *seller/buyer*'s keywords, because the result is only a difference, e.g., $M_{i,j} = M_i - M_j$, the FC cannot recover the corresponding M_i and M_j. Therefore, the proposed scheme can achieve secure fine-grained keywords comparison between tags and trapdoors.

4.4.2 Privacy Preservation of Auction Messages

The auction messages generated by *sellers* are also encrypted by the Paillier cryptosystem. They are stored in DC's database, but DC cannot decrypt it. Further, if some auction messages' corresponding keyword tags cannot match any trapdoor, they would never be decrypted by FC. Therefore, the privacy preservation of auction messages is achieved.

4.4.3 Encrypted Messages' Authentication and Data Integrity

Since the tags $C_i (i = 1, 2, \cdots)$, total trapdoors C'_{total}, and auction messages in our proposed scheme are encrypted by the Paillier cryptosystem, an adversary $\mathscr{A}$ cannot identify them, but if the adversary $\mathscr{A}$ fabricates a message and sends it to some entities, the message cannot be detected. Hence, we also sign the message by using the signature algorithm [6]. Therefore, our proposed scheme can achieve such messages' authentication and data integrity.

Table 4.1 Comparison of security level

	SESA	EFKC
Privacy preservation	✓	✓
Authentication and data integrity	✓	✓
Confidentiality	✓	✓
Secure fine-grained comparison		✓

4.4.4   Private Messages' Authentication, Data Integrity and Confidentiality

In our proposed scheme, some messages with privacy will be sent while asking for the auction array $A[]$ according to the matched tags and delivering the winner array to *buyer*. These messages are signcrypted by an identity-based signcryption cryptosystem [8]. Therefore, we can achieve authentication, data integrity, and confidentiality of the private messages.

In Table 4.1, we compare our proposed scheme with SESA [4]. We can see that both schemes achieve privacy preservation of auction messages and the corresponding keywords, authentication, data integrity, and confidentiality. However, only our scheme can achieve secure fine-grained keywords comparison. Hence, our proposed scheme can provide more security guarantees.

4.5   Performance Evaluation

In this section, we evaluate the performance of the proposed scheme in terms of computation and communication overhead. We will compare our scheme with the SESA scheme.

4.5.1   Computation Overhead

For simplicity, the cost of a pairing operation, a multiplication operation in $\mathbb{G}_1$, an exponentiation operation in $\mathbb{Z}_{n^2}$ and an exponentiation operation in $\mathbb{Z}_n$ are denoted as C_p, C_m, C_{en^2} and C_{en}, respectively. Other operations are negligible [9].

In our proposed scheme, it costs $2C_m$ to sign a message, and $2C_p$ to verify if we adopt precomputed technology. To signcrypt a message it needs $3C_m + C_p$, the corresponding decryption and verification need $C_m + 3C_p$. For *seller$_i$*, he needs $C_{en^2} + C_{en}$ to encrypt auc_i, and $(l+1)C_{en^2} + C_{en}$ to create tag C_i. Then he signs it, it costs $2 \cdot C_m$. Therefore, all *sellers'* cost is $(2 \cdot C_m + (l+2) \cdot C_{en^2} + 2 \cdot C_{en}) \cdot I$. For *buyer$_j$*, it costs $lC_{en^2} + C_{en}$ to create tag C_j', then C_{en^2} to create C_{total}'. It costs $2 \cdot C_m$ to sign it. In the end, he must decrypt and verify the signcrypted message of $win[]$, where it costs $C_m + 3 \cdot C_p$. Hence, total *buyers'* cost is $(3 \cdot C_m + (l+1) \cdot C_{en^2} + C_{en} + 3 \cdot C_p) \cdot J$ (assume J is the number of *buyers*). For DC, it must verify all messages of *sellers* and *buyers*, it needs $2(I+J) \cdot C_p$. Since homomorphic computing is a multiplication

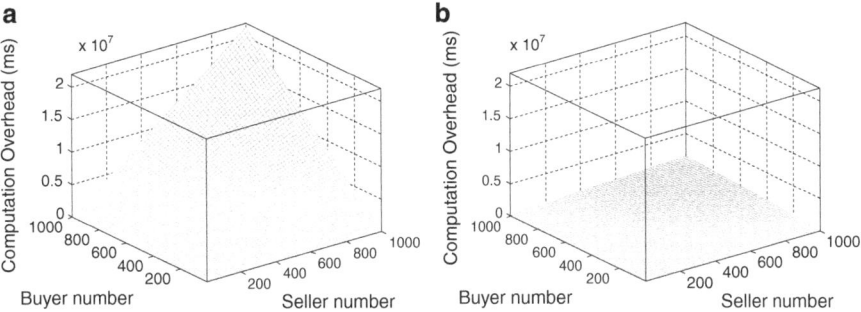

Fig. 4.2 Comparison of total computation overhead. (**a**) SESA. (**b**) EFKC

operation in $\mathbb{Z}_{n^2}$, we do not consider it. For every $buyer_j$, DC needs to sign a message $(C\|ID_{DC}\|TS)$ to FC, the signature costs $2J \cdot C_m$. And DC also receives the signcrypted messages from FC when FC needs to ask for the array $A[]$. It costs $J(C_m + 3 \cdot C_p)$ to decrypt and verify the messages. Therefore, DC's cost is $3J \cdot C_m + (2I + 5J) \cdot C_p$. For FC, it needs total $2J \cdot C_p$ to verify the messages of C, then decrypt them with $J \cdot C_{en^2}$. To ask for the pre-filtering array $A[]$, FC needs to signcrypt messages J times, it costs $J(3 \cdot C_m + C_p)$, and verifies the respond messages with $2JC_p$. Assuming that there are N tags that match the trapdoor in bid, it costs $JN \cdot C_{en^2}$ to recover auc_i $(i = 1, 2, \cdots, I)$. Finally, FC signcrypts the array $win[]$ to each $buyer$, it needs $J \cdot (3 \cdot C_m + C_p)$. Hence FC's cost is $6J \cdot C_m + (J + JN) \cdot C_{en^2} + 6J \cdot C_p$. Therefore, in our proposed EFKC, the total computation overhead is $(12J + 2I)C_m + (lJ + lI + 2J + 2I + JN)C_{en^2} + (J + 2I)C_{en} + (14J + 2I)C_p$.

In comparison, in the SESA scheme we assume it adopts the same signature, signcryption technologies and two cyclic addition groups $\mathbb{G}_1$, $\mathbb{G}_2$. EB_j makes a bid to EDR_i which costs $3 \cdot C_m + 2 \cdot C_p$, and the corresponding signature needs $2 \cdot C_m$, thus all energy buyers' cost is $5IJ \cdot C_m + 2IJ \cdot C_p$ where each EB expects to make a bid to each DER because EB cannot know which bid will be accepted; DER_i needs C_m to create a trapdoor and $2 \cdot C_m$ to sign it, at last it will decrypt and verify the message of winners which costs $C_m + 3 \cdot C_p$, therefore DER_i's cost is $4I \cdot C_m + 3I \cdot C_p$; AS needs $2 \cdot C_p$ to verify all messages—which will be $IJ + I$ times, and C_p to compare each tag—which will be IJ times, hence AS's cost is $(3IJ + 2I) \cdot C_p$; RS needs $C_m + C_p$ to decrypt a satisfied bid, therefore it costs $IN \cdot (C_m + C_p)$. In the end it will signcrypt a message of $S[]$ to DER_i, it costs $I \cdot (3 \cdot C_m + C_p)$, hence RS's cost is $(3 + N)I \cdot C_m + (1 + N)I \cdot C_p$. Therefore, in SESA, the total computation overhead is $(5IJ + 7I + IN)C_m + (5IJ + 6I + IN)C_p$.

We conducted detailed experiments on a Pentium IV 3 GHz system to study the operation cost [9, 10]. For $\mathbb{G}_1$ over FST curve, a multiplication operation in $\mathbb{G}_1$ with 161 bits, and the corresponding pairingoperation cost 1.1 and 3.1 ms. And an exponentiation operation costs 11.5 ms in $\mathbb{Z}_{n^2}$ and 2.3 ms in $\mathbb{Z}_n$. Further, we assume $N = 0.1 \times I$ in EFKC ($N = 0.1 \times J$ in SESA) and $l = 10$. As shown in Fig. 4.2, the proposed scheme greatly reduces the computation overhead.

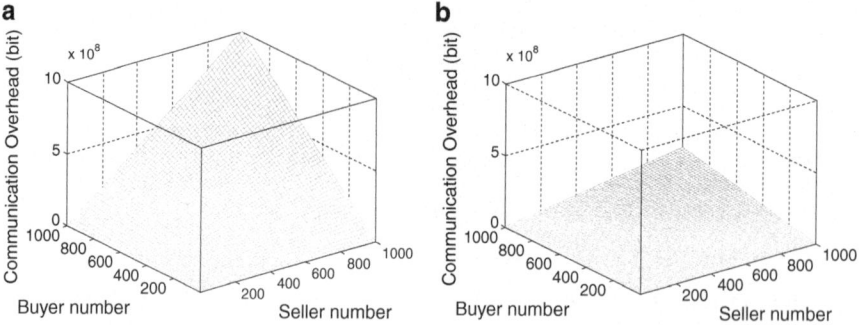

Fig. 4.3 Comparison of total communication overhead. (**a**) SESA. (**b**) EFKC

4.5.2   Communication Overhead

We divide the communication overhead of our proposed scheme into three types, $seller - DC$, $buyer - DC$, and $DC - FC$, where the delivery of winner messages are the same in SESA and our scheme, we do not compare. The message $seller$ sends to DC is formed by $(A_i||C_i||ID_{S_i}||TS||\sigma)$ where the signature includes two elements in $\mathbb{G}_1$, therefore if we choose 1024-bit $\mathbb{Z}_n^*$ and 160-bit $\mathbb{G}_1$, the total size of $seller - DC$ communication overhead is $(2048 + 2048 + |ID| + |TS| + 2 \times 161) \times I$ bits. The message of $buyer - DC$ is formed by $(C'_{total}||ID_{B_j}||TS||\sigma)$, its total size is $(2048 + |ID| + |TS| + 2 \times 161) \times J$ bits. In $DC - FC$ phase, there will be J messages of $(C||ID_{DC}||TS||\sigma)$, its size is $(2048 + |ID| + |TS| + 2 \times 161) \times J$ bits, and J messages of $(A[]||ID_{DC}||TS||\sigma)$, each $A[]$ includes average N A_i, thus its size is $(2048 \times N + |ID| + |TS| + 2 \times 161) \times J$ bits, and in $DC - FC$ there are also J signcrypted messages; its size is $(161 + 161 + |ID| + k_1) \times J$ bits (k_1 is the length of signcrypted message). In summary the communication overhead of $DC - FC$ phase is $(2048 \times (N + 1) + 2(|ID| + |TS|) + |ID| + k_1 + 6 \times 161) \times J$ bits.

In comparison, in SESA, EB-to-AS phase needs IJ messages of 963 bits, therefore the size is $963 \times IJ$ bits; DER-to-AS needs to delivery a trapdoor of 160 bits and the corresponding signature of 161×2 bits, the total size is $(160 + 2 \times 161) \times I$ bits; in AS-to-RS phase, for each DER, there are N ciphertext C_j of 160 bits and signature of 161×2 bits, hence the total size is $(160 + 2 \times 161) \times IN$ bits.

We set $|ID| + |TS|$ and $|ID| + k_1$ as 50 and 200 bits, respectively. Then the comparison of total communication overhead for SESA and EFKC are $482I + 963IJ + 482IN$ bits and $4468I + 5734J + 2048JN$ bits, respectively. As shown in Fig. 4.3, our EFKC scheme is more efficient than SESA.

4.6 Related Works

The auction market has attracted much more attention due to remarkable economic benefits in electricity trading. The corresponding issues have been extensively studied and various auction market schemes have been proposed to protect its security [3, 4]. Nguyen et al. [3] propose a demand response exchange scheme, which considers the demand response as a kind of virtual goods. This scheme focuses on transactions between users and intermediate stations. Wen et al. [4] propose a searchable encryption scheme (SESA) for auctions between energy generators and retailers. It can compare keywords secretly, and its extended scheme supports conjunctive keyword searching. However, since it adopts the hash function acting on the keywords, it can only verify whether the keywords are equal, rather than calculate the difference of keywords. And in SESA, each buyer should make a different tag message for every seller's energy he wants to bid. In this case, the computation and communication overhead are heavy.

Lu et al. [11] adopt a super increasing-sequence to aggregate all types of electricity data. In such scheme, the intermediate can achieve privacy preservation and efficiency without decrypting the received messages. Therefore, it is feasible to introduce this method into the searchable encryption auction market.

4.7 Summary

In this chapter, we have proposed an EFKC scheme for the auction market in smart grids. It can realize the secure and efficient fine-grained comparison of keywords. Security analysis demonstrates that EFKC can achieve privacy preservation of auction messages and keywords, and other traditional security requirements, e.g., authentication, data integrity, and confidentiality. Performance evaluation shows that the proposed scheme significantly improves computation and communication efficiency.

References

1. K. Moslehi and R. Kumar, "A reliability perspective of the smart grid," *IEEE Transactions on Smart Grid*, vol. 1, no. 1, pp. 57–64, 2010.
2. H. Li, R. Lu, L. Zhou, B. Yang, and X. Shen, "An efficient merkle tree based authentication scheme for smart grid," *IEEE Systems Journal*, http://ieeexplore.ieee.org/xpls/abs_all.jsp?arnumber=6563123.
3. D. Nguyen, M. Negnevitsky, and M. de Groot, "Pool-based demand response exchange—concept and modeling," *IEEE Transactions on ower Systems*, vol. 26, no. 3, pp. 1677–1685, 2011.
4. M. Wen, R. Lu, J. Lei, H. Li, X. Liang, and X. Shen, "Sesa:an efficient searchable encryption scheme for auction in emerging smart grid marketing," *Security and Communication Networks*, http://onlinelibrary.wiley.com/doi/10.1002/sec.699/full, 2013.

5. D. Kang, B. Kim, and D. Hur, "Supplier bidding strategy based on non-cooperative game theory concepts in single auction power pools," *Electric power systems research*, vol. 77, no. 5, pp. 630–636, 2007.
6. B. Libert and J. Quisquater, "The exact security of an identity based signature and its applications," *Preprint available at* http://eprint.iacr.org/2004/102, 2004.
7. L. Chen and J. Malone-Lee, "Improved identity-based signcryption," *Public Key Cryptography-PKC 2005*, pp. 362–379, 2005.
8. P. Barreto, B. Libert, N. McCullagh, and J. Quisquater, "Efficient and provably-secure identity-based signatures and signcryption from bilinear maps," *Advances in Cryptology-ASIACRYPT 2005*, pp. 515–532, 2005.
9. "Multiprecision integer and rational arithmetic c/c++library," http://www.shamus.ie/.
10. M. Scott, "Efficient implementation of crytographic pairings," *Available:*http://ecrypt-ss07.rhul.ac.uk/Slides/Thursday/mscott-samos07.pdf.
11. R. Lu, X. Liang, X. Li, X. Lin, and X. Shen, "Eppa: An efficient and privacy preserving aggregation scheme for secure smart grid communications," *IEEE Transactions on Parallel and Distributed Systems*, vol. 23, no. 9, pp. 1621–1631, 2012.

Chapter 5
Conclusions and Future Directions

5.1 Research Conclusions

The objective of this book is to present the state-of-the-art solutions to secure and privacy-preserving communications in smart grids. Specifically, we obtain the following important insights:

- *Demand response should not only provide privacy preservation of electricity demand but also mitigate the damage caused by the exposure of secret keys stored on smart meters.* In the demand response of smart grids, adversaries might eavesdrop on the communication between users and control center as well as identify the users' electricity demand. With this information, they are able to track and learn about the users' habits or lifestyles. Moreover, adversaries might compromise the smart meters and further obtain stored secret information such as their session keys and private keys. Both privacy preservation of electricity demand and keys exposure must be considered and captured in the demand response schemes.

- *A lightweight authentication is essential to computation-constrained smart meters.* The smart meter is vulnerable to malicious operations. It is indispensable for electricity utility to prevent malicious operations. If the attacks cannot be detected, the control center will be misled and make incorrect decisions, such as sending a false pricing information to the customers. Therefore, it is extremely important to develop an authentication scheme to detect the replayed/injected messages. In addition, a smart meter is only equipped with limited resources. Accordingly, the authentication scheme should be lightweight WLAN bandwidth whenever available.

- *Energy auction needs to consider privacy preservation and fine-grained keywords comparison.* Firstly, due to the confidentiality of the auction information, privacy preservation is extremely important. In addition, fine-grained keywords comparison is significantly useful. For example, with the difference of price keywords, energy buyers can filter out the energy with

H. Li, *Enabling Secure and Privacy Preserving Communications in Smart Grids*, SpringerBriefs in Computer Science, DOI 10.1007/978-3-319-04945-8_5, © The Author(s) 2014

reasonable price. As such, both privacy preservation of auction information and fine-grained keywords comparison should be addressed in energy auction schemes.

5.2 Future Extensions

The solutions reviewed in this book cover privacy-preserving demand response, lightweight authentication, and fine-grained keywords comparison. However, due to the complexity of smart grids, there are many open issues to extend the research in the following directions:

- *Fine-grained multi-keyword search can be further considered in energy auction schemes.* In an energy auction scenario, as outsourced energy auction data may contain sensitive privacy information, they need to be encrypted before being transmitted to the auction center. This severely limits the usability of data due to the difficulty of searching over the encrypted data. To address the issue, searchable encryption technology has been proposed in literature as a fundamental approach to enable keyword search over encrypted data. Existing searchable encryption schemes can achieve fuzzy keyword search, ranked keyword search, and multi-keyword search, etc. It is, however, a challenging issue to develop more complicated and comprehensive search functionalities.
- *It is challenging to develop multi-authority access control with efficient attribute revocation in smart grids.* In smart grids, the control center collects and aggregates users' electricity data via smart meters. The aggregated data is also of great use for markets. To efficiently and securely distribute these data to markets, the existing schemes use the Attribute-based Encryption (ABE) technique to achieve privacy preservation of sensitive data and fine-grained access control. However, the efficient attribute revocation problem has not been studied well. Current solutions to attribute revocation are not efficient since they have to update all the ciphertexts, which contain the revoked attribute and send them to each non-revoked user.